Arthur Clayden

The revolt of the field

a sketch of the rise and progress of the movement among the agricultural labourers

Arthur Clayden

The revolt of the field
a sketch of the rise and progress of the movement among the agricultural labourers

ISBN/EAN: 9783744745284

Printed in Europe, USA, Canada, Australia, Japan

Cover: Foto ©berggeist007 / pixelio.de

More available books at **www.hansebooks.com**

THE

REVOLT OF THE FIELD:

A SKETCH

OF THE RISE AND PROGRESS OF THE
MOVEMENT AMONG THE AGRICULTURAL LABOURERS,

KNOWN AS THE

"NATIONAL AGRICULTURAL LABOURERS' UNION;"

WITH A

Reprint of the Correspondence to the DAILY NEWS
DURING A TOUR THROUGH CANADA WITH MR. ARCH.

BY

ARTHUR CLAYDEN.

London:
HODDER AND STOUGHTON,
PATERNOSTER ROW.
MDCCCLXXIV.

TO SAMUEL MORLEY, ESQ., M.P.

In no spirit of sycophancy, but as a trifling expression of my admiration of your princely generosity in contributing the sum of five hundred pounds to the National Agricultural Labourers' Union, when its early struggles and its gigantic difficulties made the gift of fourfold value, I dedicate this brief and imperfect record of its proceedings during the two years of its existence. That gift, with the moral force attending it, and your presidency of the meeting in Exeter Hall, on the 10th of December, 1872, determined in a great measure the fate of the movement; and I doubt not the labouring classes, not merely of the important city which you represent, but of the whole country, and I may say of the whole civilised world—for I have heard an enthusiastic acknowledgment of the service from thousands of working men in the foremost city of America—will hold the splendid illustration of true sympathy with the honest toiler in lasting and grateful remembrance.

A. C.

PREFACE.

THE question of the condition of our agricultural labourers having become one of universal interest, through the efforts put forth by Mr. Arch and his coadjutors, it has been thought desirable to place before the public the means of forming an estimate of the true character of the movement of which he is the presiding genius. Although the writer, through an early identification with the rural awakening, and a later co-operation as a member of the Consulting Committee of the Union, is able to furnish tolerably accurate data for the dispassionate outside inquirer's judgment, he desires it to be distinctly understood that he has written in no official capacity, but entirely on his own responsibility. And the same remark applies to the series of letters appended, which appeared in the columns of the "Daily News" during the writer's travels with Mr. Arch through Canada. Mr. Arch is in no way responsible for any of the statements contained therein.

FARINGDON, BERKS,
February, 1874.

CONTENTS.

	PAGE
General condition of the Agricultural Labourers anterior to the Union—Efforts of Canon Girdlestone and others to ameliorate the same ...	1
Joseph Arch—His birthplace, early history, and call to his life-mission ...	6
First great Labourers' Meeting at Wellesbourne ...	9
Inauguration of the Union at Leamington...	21
Rules of the Union, and Introductory Address by Mr. Arch...	23
Incidents of the Movement—a Lock-out ...	31
Important Trial at Faringdon to test the right of Public Meeting—Mr. Fitzjames Stephen, Q.C., and Mr. Edward Jenkins, M.P., in defence of Mr. Arch and others—Splendid Victory...	37
Great Meeting in Exeter Hall ...	44
"Labourers' Chronicle" Manifesto ...	46
Speeches of Union Delegates ...	49
Great Out-door Meeting at Newbury—Speech of Mr. Arch ...	75
A Workhouse Scene ...	87
The Guardians and the Labourers ...	91
A Cockney among the Rustics—The Labourer interviewed by a "Daily Telegraph" Reporter ...	95
Popular Ballads of the Union ...	99

CONTENTS.

	PAGE
"Punch" and the Labourers' Enfranchisement	115
The Labourer and the Land—Mr. Arch's pet project	117
The Labourer and the Church — Theory of the general antagonism	125
Chipping Norton Prosecution of Labourers' Wives	131
Second Annual Conference at Leamington	145
Altercation between Messrs. W. G. Ward and Edward Jenkins over article in "Chronicle," "The Battle and the Victory"	145
Public Meeting—Speeches of Mr. Dixon, M.P., and others	155
The "Times" on the political aspect of the movement	181
The Enfranchisement of the Labourers — Speeches of Mr. Trevelyan and others in the House of Commons	182
With Mr. Arch in Canada	201

THE REVOLT OF THE FIELD.

THE year 1872 will be ever memorable as the year of the birth of the great movement among the English agricultural labourers. Other premonitory stirrings in the rural districts had for some years past been occasionally heard of. A clergyman of the Church of England—Canon Girdlestone—had obtained considerable notoriety and endured no little reproach on account of efforts which he had been led to put forth on behalf of the labourers in the fields. As the State-appointed parish priest of an agricultural district, he had been brought into close contact with that social wretchedness which has now become pretty well known to the whole civilised world. And, unlike too many others, he was not content to acquiesce in the state of things as an ordinance of God. The universal mode of dealing with the intense poverty of the rural swains by the multiplication of small charities, to be supplemented by the legal provisions of the Poor Law Board, did not satisfy this Christian pastor. He saw that the remedy was almost worse than the disease, superadding to the evils of poverty the curse of pauperism. Hence his practical efforts in the direction of migration. Coming from a part of the country where labour was more scarce and better remunerated, the now familiar idea of raising the tone of the village labour market by drafting off the surplus hands to other localities occurred to him. And by this means, and the free use of the press in repeated and most forcible appeals to the public, a considerable impression was made on the public mind. The subject was talked about in various sections of society, and occasionally the walls of St. Stephen's echoed to the notes of warning which were addressed by honourable members of Radical proclivities to the landowners on the Conservative benches.

And considerable amelioration undoubtedly had taken place in the condition of the English field workers. Their homes had long engaged the thoughtful attention of philanthropists. The present Earl of Shaftesbury, in the course of a debate in the House of Commons some quarter of a century ago, had used such words as these respecting the cottages of our peasants:—
"There is no such thing as a home, and the man who has a wife and children is not the head of a family, but the chief pig in a pig-sty." Landlords in various localities had anticipated the present movement by timely concessions to public opinion, and, I will in justice to them add, by an honest endeavour to rise

to the full height of their vast responsibilities. In the royal county of Berkshire the late "Radical Peer"—the Earl of Radnor—had at a great cost rebuilt the village of Coleshill, erecting for the labourers on his estate homes replete with comforts, of most tasteful elevations, and surrounded with adequate garden-ground. Another extensive landowner—Mr. Henry Tucker, of Bourton House—threw himself into the matter with considerable energy. At an anniversary meeting of the Faringdon Agricultural Library in the year 1860 he furnished the members with accurate statistical information as to the state of the cottages in the several parishes comprising the Faringdon Union; and within a short period afterwards he wrote a letter to the "Times," embodying therein some of the more startling facts previously communicated to the meeting at Faringdon. In the hope of still more effectually promoting the laudable object he had in view, Mr. Tucker, in the spring of 1862, offered a prize of fifty guineas "for the best essay upon the following subjects, relative to the dwellings of the labourers in the agricultural districts of England:— 1. Tracing and explaining the cause of their present crowded and defective condition, with any authentic information illustrative of that condition. 2. The effects, moral and physical, which such condition has upon the inmates. 3. Suggestions as to the best practical means of ameliorating the evils, either by compulsory legislation or otherwise." Mr. John Walter, the Liberal member for Berkshire, undertook to examine the essays, and has himself evinced a laudable anxiety in the same direction in his own neighbourhood.

Various isolated attempts had also been made in different parts of England to bring about a better state of things among the agricultural labourers, but without much result. An interest in the subject was awakened in some quarters, and sundry legislative enactments were entered upon the statute-books; but to all intents and purposes the general condition of the men was very slightly improved. Their utter prostration, and the immense power of their employers, backed up by the social and political influence of the wealthiest aristocracy in the world, seemed to make their case hopeless. Benevolent and philanthropic men on every hand looked on their position and sighed. Contrasting the plenty and comfort of their own homes with the squalor and wretchedness of the labourers', they had often and long looked about them for the key to a reform. To ask men with ten shillings a week to live upon, and nearly as many mouths to feed and backs to clothe out of it, to lay by for a rainy day was cruel mockery. To advise them to ask for more was to declare war against your friends and neighbours. So the thing seemed destined to remain as it had been from the beginning. Legislation was powerless and benevolence was at sea. The prophetic words of

Goldsmith bid fair to receive an illustration other than that of the skilful engraver :—

"Ill fares the land, to threatening woes a prey,
Where wealth accumulates and men decay."

Such was the state of things at the commencement of the year 1872. A thousand earnest men stood baffled in view of a social blot which they could neither remove nor endure; and, as has often happened in human experience, "man's extremity was God's opportunity."

Within about four miles of the genteel town of Leamington is the quiet and picturesque village of Barford. The mainstay of the place are sundry well-to-do maiden ladies, who, from choice or circumstance, have made it their home. When it is remembered that within an hour's ride are the grand old historic castle of the King Maker, the venerable and unique ruins of Kenilworth, and the classic shrine of Stratford-on-Avon, it will be at once seen that these respectable and well dowered ladies have not greatly erred in their selection of a dwelling-place. But others reside in this romantic spot whose lot has been somewhat different from these ladies of fortune and the numerous

"Heirs of flattery, to titles born
And reputation and luxurious life."

Toiling men and women have their homes in this Warwickshire village. Right in its centre stands a comfortably-built cottage. If we had opened the door, say, two years ago, we should have found a neatly-dressed labourer's wife ironing her husband's shirt. That husband was probably some twenty miles off, cutting and putting in order Farmer Brown's quickset hedge. By-and-by, as it was Saturday, her "man" and the lad would be at home; as he generally managed to be at home for the Sunday.

As you take the proffered chair and look about you, you see sundry signs of its being a religious household. The Bible is at hand and looks well used. The few books on the shelf are *not* exactly shilling novels. The coloured prints on the walls savour more of Palestine than the prize ring, and to confirm your impressions, a Primitive Methodist preacher's plan is seen behind the inner door. Yes, this is the home of JOSEPH ARCH— a name soon to be known wherever the English tongue is spoken. From the age of nine years he has been a farm labourer. In this village, on the 10th November, 1826, he was born. His father had been an agricultural labourer, and, after fifty years of toil, had been only saved from the much-dreaded indignities of the workhouse in his old age by the manly determination of the son.* "With a sigh the old man crept into bed and wept, knowing he was, after all his work, to become a burden to those he loved, and who, as he knew, had

---

* "Life of Joseph Arch." By F. S. Attenborough. (Vincent, *Leamington*.) P. 37.

barely enough for themselves." "I be afeard, Joe, the parish ull give thee nothin' for me, be'n as yer a Dissenter." Joe was not anxious that they should; but Joe's wife had been in the habit of earning a couple of shillings a week at charing, and now that the old man wanted nursing, she had to give this up and stop at home. To the guardians Arch made a reasonable offer. "Gentlemen, I don't want you to support my aged father; but if you will give my wife one shilling and sixpence towards nursing him, now that she is cut off her charing, I shall be much obliged to you. It isn't much; it's less than the loss of my wife's earnings, and nothing towards the expense." "Certainly not, Arch; your father can go to 'the house,' and you must pay one shilling and sixpence towards the expense." "Good morning, gentlemen. I'd sooner rot under a hedge than he should go there."

Such was the man whom Providence had selected for the great work of raising up a pauperised and downtrodden class of English workmen. Generation after generation of these valuable citizens — these tillers of the soil, these producers of the country's wealth, and, in the hour of danger, defenders of that wealth and guardians of our shores—had lived their hard and toilsome life, and passed away through the workhouse to a pauper's grave, beneath the eye of the parish squire and priest. And only here and there had there been one to dispute the morality of the system. With the complacency of the priest and Levite of old, social well-to-doism in our villages, both lay and clerical, Church and Dissent, had settled down into a state of acquiescence in the "Divine appointment;" and as Hodge lay there by the wayside, maimed and bruised by the cruel exigencies of his lot, they had been content to pass by on the other side.

But it had at length pleased Almighty God to raise them up a deliverer, and, as in the case of the oppressed toilers of old—the agricultural labourers of the Mosaic era—" one from among themselves."

On the 7th February, 1872, the first open-air meeting of labourers was held in the village of Wellesbourne. On the 13th another meeting was held in the same place under a fine old chestnut tree, which will henceforth be numbered amongst the historic trees of England. At this meeting Joseph Arch made his maiden speech; and our "world-wide" circulating journal, the "Daily News," with that unerring instinct which has been the secret of its splendid success, at once appreciated the significancy of the movement, and, through the agency of a gentleman who had, by his graphic pictures of the great Franco-German war, 'made' both himself and his paper, at once published tidings of the revolt to the world.

As this meeting, like the small, narrow stream in Gloucestershire which widens as it flows until it becomes the magnificent Thames of London, must ever possess a great interest to Englishmen, I will quote a description of it from a life of Joseph Arch written by the Rev. F. S. Attenborough, a Congregational minister of Leamington.

A LABOURERS' MEETING.

An English village on a bright summer day is a sight worth seeing. The straggling street; the great leafy trees; the old grey church with its green hillocks, among which the sheep feed; the pond, with a tired horse or two lounging on its brink and a score of ducks dabbling in its muddy waters; and the quaint gabled cottages, set round with old-fashioned English flowers, such as stocks, primroses, hollyhocks and snapdragons, covered with dark mossy thatches, and crowned with grotesque chimneys, which only stand by leaning upon one another—all combine to make one of the homeliest and most picturesque scenes to be found in the world. No doubt there are details in the picture to which the fastidious might fairly take exception. Many of the cottages are small, dark, and damp—very pretty in a water-colour, but very wretched to eat, drink, live, and sleep in. The pavements look like organised contrivances to effect the ruin of human feet and shins; the open gutters are a trifle too aromatic; the manure heaps are more frequent and fragrant than pleasant; and certain needful structures, which we forbear to name, are so near to the wells as to suggest the idea that many of the villagers drink their water mixed with ingredients which are not usually considered as being either good for food or pleasant to the eyes. Having regard to picturesque effect, we may pronounce most English villages to be successes; but having regard to light, air, pure water, cleanliness, decency, and health, nine out of ten of them must be pronounced failures. But none the less, our villages, in the summer, under the sun which glorifies everything on which he shines, are really very pleasant to look upon.

They are not so attractive in winter. The trees, stripped of all but a few withered leaves, stand trembling in the cold, like persons who have seen better days; the gardens are only so many brown earth-blots; the pavements are full of treacherous puddles, which bode ill to other than water-tights; and the street is covered with a layer of greasy clay, which cakes on the boots and gaiters of the men, till they can hardly move in them or get out of them, and makes the poor horses slip hither and thither as though they were learning to slide, or had been drinking as much brown beer as their drowsy teamster.

In the village the writer is thinking about the ills and pains of winter were yet lingering. The short February day was dying out; through the cottage windows the faint light of the lean "short sixteens" was beginning to glimmer, and the men, moist, cold, and weary, were getting home. To-night they don't straggle in one or two at a time, but come in groups of eight or nine or more. They march quicker than usual, their air is brighter, their faces are more occupied, and it is plain they have something special on hand. Presently they come out of their cottages cleaned and brushed up a bit, and make for an open space at the end of the village. Three or four men drag an old waggon into the centre; all look and wait, and there soon rises that buzzing murmur caused by the motion and questioning of a crowd. A team comes creaking and jolting down the street, and pulling short up opposite the concourse, the driver, a rough, hard fellow from another village, wants to know "what's oop at Grindington to-noight?" "Whoy, doant e know as Joe Arch is a cummin oaver?" "Noa, e ant erd, but if 'e be, dang it, if e doant toi oop th' team and wait a bit!" Hark! Tramp, tramp, tramp, comes through the thick air; a hoarse cheer is raised, and some thirty or forty fellows from another village emerge out of the darkness and blend with the crowd. Another detachment arrives, and then another, until not fewer than five or six hundred men are assembled. It is evident they mean business. For a group of labourers they are marvellously quiet and self-possessed; and looked at by such light as a few lanterns, a lamp or two, and some bottled candles cast upon them, their faces show settled determination, and something which looks like hope, and is, in many a case, there for the very first time. Two or three farmers hanging on the rear of the crowd eye it with manifest uneasiness. A little stir is excited by the arrival of a cart containing three or four men, who soon begin to arrange certain note-books and pencils, and by this are declared to be reporters. A cottage door opens on the right, and from out the warm light it emits come five or six labourers, who make straight for the waggon. The crowd divides to let them pass, and as those nearest recognise one of the number, they break out into lusty "cheers for Arch and the Union," which cheers are taken up by the crowd, and repeated again and again, until they ring through the village, waking the children, rousing the very jackdaws in the steeple, and filling the two old maids who live at "Verbena Cottage" with the dread foreboding that their time has come. An aged man rises in the waggon; the cheers cease, and amid perfect silence he begins to speak. He is manifestly new to the position and to the work, and at first proceeds with hesitation and discomfort. His "old ooman" is listening to him, so is his master, and he stands in awe of both. As he goes on his courage strengthens. The pluck which in our Anglo-Saxon race always rises to the surface in the time of need rises in him, and now, regardless of everything and everybody, he gives vent to his deep convictions and long-pent-up feelings in plain, homely, fervid speech, which compels attention and wins unmistakable assent from those who hear it. His is a strange say for this land and these times! England is a rich country, the richest in all the earth; but this man, one of England's sons, and one of her most deserving ones, tells of a poverty that is most woful. This is a free country,—our songs say so, our Parliament men say so, our newspapers say so; but this man tells a tale of oppression and servitude which crimsons the honest cheek with shame. This is a contented country; our peasants live in sylvan cottages, they serve considerate masters, they are well nurtured, their lives pass smoothly on, their old age is calm, unanxious, and respected. So we have heard and read. But this man tells a tale of misery and discontent and despair which scatters all these dreams and pictures to the winds and fills one with wonder. Listen! He is seventy years of age. He has been a farm labourer sixty-three years, having begun when he was barely seven. He has preserved his character; he has worked hard; except in times of sickness, he has kept off the parish; his master and the squire and the parson give him a good word, and smile benevolently when he makes his bow to them, as "he is in duty bound to do." He has given sons and daughters to the State. Of these, some died in infancy, because, as he says, he couldn't properly nourish them. The low fevers which hang about ill-drained, ill-lit houses seized them, and they died; died because they were not and could not be sufficiently fed. Others pulled through, and grew up, he knows not how. He could not educate them, for the schools were few and poor; and then he hadn't the pence for the fees. True, their mother had "a little larnin', but she was moastly at 'th' House,' or else i' the fields, helpin' to urn a bit o' bread; and when she got whoam she wur tired; and for the matter o' that, so wur the children, for all on 'em as could had been a bird scarin', or stonin' i' the lond, or lookin' for a bit o' firewood ; fur ye see the family wur large, and things wur dear, and when we'd paid for our flour, wi' all we could do, we had but a moite left." But the children grew up. Of the girls, two or three are the wives of labourers : they have families, are poor, and can do nothing to help their aged relatives on either side. One, the youngest, is in service, "and being a raul good lass, contrives to remember the old folks at home." Of the lads, the bones of one are whitening in India, where he fell ; another sleeps with his comrades in the English cemetery at the Crimea ; a third was lost at sea ; one or two yet remain, and they are on the soil. This man deserves well of his country ; and the country recognises his deserts to the extent of paying him eleven shillings a week, stipulating that he shall work for this from dawn to dark. He hears of privileges and such like things, but he knows that for the last twenty years, taking one week with another, "he ant earned so good as eleven shillin'." He is now old, his joints are stiff, his strength is departing, and very soon he will have to stand by. What awaits him then? "A bit o' stone-breakin' on the roads, two shillin' a week from the parish, and a loaf or two." He will be a burden to the guardians, the guardians will let him know it, the poor old fellow will smart under the sense of it, will wish he was dead,—so will the guardians,—and before long both will be gratified. Poor old man ! He has always been insolvent, he has never been half fed, or content, or free ; he has never felt himself to be a man, and knows that others have regarded him as a mere farm machine.

"He reads in the paapers as 'ow the country's growin' richer and richer ; as 'ow the cairpenters and such, by union, have bettered themselves, and come to have a share in the gen'ral improvement, and thinks summut should be doin' for th' labrin' class. He has no hope for himself, but afore he dies he'd uncommon like to see the young uns doin' better nor he's done. He's heerd o' the Wellesbourne men, and Muster Arch, and seein' as th' master and th' squoire canna graetly harm him, sin' he's nigh done, he's made bold to ax Muster Arch to come and tell em down here what they must do to get a trifle more wage and a bit better food; and Muster Arch, like a good un as he is, has come, and will speak to um himself." Tremendous cheering, amid which Joseph Arch rises in the waggon and faces the crowd. Move the light a little nearer that we may take a good look at this man! Of late he has been holding meetings of this character in his own county of Warwickshire, and he means to hold them, as far as he can, all through England; advocating as his end, shorter hours and better pay, and as his means, union.

These men before him, and all of their class, have any quantity of faith in him. They call him "our man, our Joe, the labourers' hope, apostle, friend." His influence over them is unbounded. He has but to speak and it will be done. If he likes to retaliate on the farmers, as they seem inclined to do on him and his men, there will be mischief—mischief for which the farmers will be solely to blame. He has but to urge vengeance, and night after night flaming stacks will illumine the darkness, and the whole country will be laid waste. Be very careful, Joseph Arch! You have at this moment more power than any man in England over the most downtrodden and oppressed class of the community. They are smarting under the sense of injury and injustice. You hold them as men hold hounds in a leash,—hold them with a judicious hand! If you loose them and let them go, they will, without a doubt, spring as you direct them, but not even by you will they be called back. What will this man, so suddenly yet so certainly possessed of such vast confidence and influence,—what will he do ! Must we hope, or must we fear? Look at him as he stands there eyeing the crowd, the light full upon his face. What sort of a face is it? Subtle, hard, selfish? Is the forehead low and dark, or contracted and retreating? Are the eyes closely set, furtive, and vicious? Is the chin long, sharp, and cruel? If so, fear the worst. But these features are not here; I shall hope in a man who has this face, and in his work. Brow broad and open ; eyes frank and brave ; chin square and firm ; face browned with exposure and marked with small-pox, but thoroughly honest and manful ; head round, very well set, and always erect, except before God. I can trust this man. I am sure he is true. He means fair play. He hates oppression, whether it is on the side of master or on the side of man. He will wilfully injure no one. He will stand like a wall in defence of right or in opposition to wrong. This man has roused in me the respect and enthusiasm he seems to excite in all who come to know him. You may laugh at me, but I am sure if Joseph Arch had been other than peasant-born, and had carried into other, and, socially, higher spheres the same fearless, strong, wise nature he now possesses, his country would have heard good things of him. In Parliament, Bright and Cobden would have found him a trusty henchman and ally. In the Navy of the past, Nelson would have known him as a man in no wise likely to disappoint England's proud expectation. In war, the Iron Duke would have cast on him that curt smile which a few yet remember, and would have sent home some pithy word about his truth and valour. But let us hear him. He draws his frame up to its full height of five feet eight-and-a-half or nine, throws his deep chest well forward, and as his face lights up with a singularly intelligent and kindly expression, in a clear, ringing voice, which reaches the remotest listener, he utters such sentiments as these :—

"I shall perhaps do right at the outset if I tell you I am a working man—one of yourselves. It has been insinuated by some persons round my neighbourhood, that if some gentleman had taken up this question they should have had no doubt whatever of its success. I do not happen to be one of those who think like that. I think if any one is able to understand the wants, inconveniences, and wishes of the farm labourers, it is that man who has been a farm labourer for over twenty years, even though he may not have been educated as a gentleman. I have had such experience; but I must tell you, before I proceed further, that you must not expect from me anything like oratory. I was sent to work, like most of your boys, when nine years of age, and from then till now I have had to labour with my hands to procure bread. I have been able, by the help of Divine Providence, by hard work—which I don't dislike—and by much sacrifice, to bring up a family of seven children; therefore I have had some little practical knowledge of the trials and troubles of the labouring man. And now I would just say here that I do not want, in anything I shall say to-night, to cause a bad feeling between the farmer and his men. I am not here to advocate ill-feeling, rowdyism, or anything like that. I am not here to advocate an onslaught on your employer's property, nor that you should jeopardise his rights and interests in order to extort from him an advance of wages. We are going to burn no farmer's ricks, to rob no farmer's house, nor kill any farmer's cattle,—are we? I am come here to-night, by request of your Committee, to explain to you, in my simple way, as a working man, wha is the great necessity of the day with regard to the working man. There is a great, deep necessity, which must soon be met. I know of efficient, able-bodied men who only get 10s. per week. Lately I went to the house of an old experienced shepherd, whom I found living with his wife in a hovel not fit to turn pigs in, and the poor old man, who works from five in the morning till seven or eight at night, only receives 5s. a week and two loaves. Only yesterday, a waggoner told me he works from four in the morning till eight at night, except on Sundays, when he works from five till six. The man works one hundred hours in the seven days for 12s. 6d., or 1½d. an hour. I want to ask every sensible individual how it is possible for a man with four or five children to support them decently, respectably, and properly on 12s. a week. If there is any man here who can tell us how a family can be brought up as it should be on 12s. a week, we shall all be very pleased indeed to hear him. The working man needs more money, and must have more. It is very evident that, whatever some part of society may think of us, each individual Englishman has responsibilities of his own, whether you find him in the House of Peers, in the counting-house, in the market, or in the field — every individual man has responsibilities resting upon his shoulders, which he must carry out more or less for himself. We have our responsibilities; but the question is—Are we able to carry them out on 10s. or 12s. per week? The tradesman has the means of enlightening and instructing his children, and rendering them respectable and useful members of society; but the agricultural labourer is deprived of such a privilege. There is no labourer in this meeting who can feel, when he looks upon his family, that he has done all for them that he should like to have done, or even that society expected at his hands. Why is there so much ignorance amongst us? Is it because parents are not careful enough to send their children to school? I admit there are some fathers amongst us who are a little bit careless in this matter; but the fact is, that the exigencies of existence demand that the boy should go to work at eight, and keep on till he is eighty, if his strength holds out so long.

How is it possible for the father to send his children to school when he cannot maintain them? It matters not how cheap the education is if you cannot spare the child's labour, to enable him to go to school. How can we raise our children to a higher level, how can we educate them, except by an increase of wages? Then, again, the working man has been called a spendthrift; it is said he is always in debt; and some set him down as a positive scamp. Suppose a man by the strictest economy is enabled to keep out of debt on 12s. a week, how can he help getting behind in his payments when illness overtakes him or any of his family require nourishment and medicine? He is unprepared for the least adversity; and when once behind how is it possible he can recover himself, when at the best he only kept out of debt by the skin of his teeth? It may be said there are charities for the poor man in his trouble. No doubt these are very good things in their way, but I can hardly say that I think them to be quite so good as some people try to make them out to be. I am an Englishman, and I know that I am speaking to an audience of Englishmen. There is not a man here to-night, however poor, but he likes to wear his own coat, and to cut his own loaf; and if a man can only have a red-herring for dinner, he likes to pay for it. Just let us look at this charity a bit. The parson of the parish is generally the person who is the trustee or dispenser of the charities, and as he comes round and distributes them, he expects us to be very grateful, thankful, and obedient. If you assert your manhood, and say, "I am a man as well as you," he will not give you much charity. I think the wisest step to take in the matter is to claim what is right for our labour, so that we can buy our own comforts, and tell these kind respectable people to keep their broth at home. I want to see my fellow-workmen raised from serfdom and slavery. Why, have we not as much right to have money and pay for our own clothes as the great Premier of the country has, and as the farmer has? I maintain that we have. Let us ignore this state of slavery, and get from under the hand of tyranny. Talk of charity! Let us claim our rights with English independence, honesty, and manhood.

"It is said by some, that capitalists have decidedly the worst side of the question—are the worst off. I really cannot believe it. I remember some few years ago a farmer came into my own neighbourhood and entered upon a farm with a capital of £200, as he told me himself. Three or four years after, he asked me to come and assist him in the hayfield. During the four years he had bought a fine hunting horse, and whereas at first he used to work in the fields himself, he had latterly left it all to others. As I was working, he came up and said, 'I have been losing money ever since I came on the farm.' That puzzled me, and I said, 'Why, you told me soon after you came that you had only £200 when you took the farm. I should judge that the hunter you have bought is worth £150. Now if you commenced business with only £200, how is it that you could buy that hunter? How is it that you are so much more the gentleman now than you used to be, if you have been losing money all the while? The man was silent. I assert, fearless of contradiction, that the capitalist has decidedly the best of it. When I said at Tachbrooke, last week, that the labourers were a downtrodden class, it seems I offended some people, who, I am told, say it is a barefaced falsehood. I ask you, Is it true? A man who publicly said it last Sunday is paid to preach to the people and save their souls. Can *he* live upon 12s. a week?—and if he can't, a bachelor curate, how can we family men? I value my character as much as he does his, and I hurl back the imputation, and say it is he who is deceiving the people, and not me. I want us all to be honest, sober, and free, like

THE REVOLT OF THE FIELD.

men have God on our side, right on our side, public opinion on our side, and law at our back. Shall we remain in serfdom? Shall we see our wives suffer for want of proper nourishment, and our children grow up in ignorance, and not try to help ourselves out of this wretched state? UNION is our only hope. Let us stand shoulder to shoulder and go forward with confidence. Ask for what is fair, and when you have asked it stand by it at all costs. Don't compromise and don't be intimidated; don't look at the toes of your boots, but look at your master right in the face, as honest men. Let no sophistry, no bribe, no threat, shake your resolution or disunite your ranks. Stick together, and the day of your emancipation is at your own command. I move '*That this meeting thoroughly approves the Agricultural Labourers' Union, and pledges itself to do all that is in its power to support its interests.*'"

Enthusiastic cheering, giving in of names and subscriptions—and the assembly breaks up. This is a sample of the kind of meeting which has been held night after night since February last, and of the style of Joseph Arch's addresses. His words are always direct, temperate, manly, and thoroughly honest. The labourers could not have a better champion, nor the farmers a fairer opponent. He is worth knowing. Labourers should know and honour him for his wise advocacy of their most deserving cause. Farmers should know and respect him for his burning denunciation of all violence and coercion, for the moderation of his counsels, and for the fairness of his demands. Society should know him and respect him as a man who is wisely and successfully heading the greatest movement of these times.

This meeting soon bore fruit. On the 11th March two hundred labourers of Wellesbourne resolved to make an effort for an increased wage; and being unable to obtain it as they wished in a peaceful and businesslike manner, they resolved to "strike." Thanks to the publicity given to the movement by the press, a considerable amount of sympathy was soon evoked, and on an appeal being made to the trades' unions throughout the country, funds soon began to pour in. An extensive migration of labourers to the north was initiated, and on the 29th of March the inaugural meeting of the Warwickshire Agricultural Labourers' Union was held at Leamington. At this important and most novel meeting the Hon. Auberon Herbert, M.P., presided, and letters of encouragement from several other members of Parliament were read.

A donation of one hundred pounds was handed in from "a friend of the movement," who wrote, "The right to combine must be fought for to the death"—a sentiment which was received by the assembled workmen with enthusiastic and prolonged cheering.

The immediate result of this meeting was a general rise in wages in the district, and in utter ignorance of whereunto the thing was going to grow, Joseph Arch meditated a return to his usual work. But it was not to be. As the "woman of Samaria" had, under a mightier inspiration, to leave "her water-pot" to make known her wondrous revelation to the "men of the city," so this man must put down his hedge-hook and his spade, and go forth throughout the rural districts of England with his message of hope and deliverance.

The "fulness of time" for the emancipation from social thraldrom of the half-million of field workers had come, and with the hour had come also the man.

On the 29th May a conference of agricultural delegates representing some twenty-six counties was held at Leamington, and Mr. George Dixon, one of the members of Parliament for Birmingham, presided.

At this meeting the "Warwickshire" Agricultural Labourers' Union was merged in the "National," and the foundations were laid of the present widespread and most potent agitation. An executive committee of twelve *bonâ-fide* agricultural labourers was formed, of which Joseph Arch was made the chairman. The names of the twelve committee-men were as follows:— E. Russell, E. Pill, G. Allington, T. Parker, T. Biddle, J. Prickett, J. Harris, E. Haynes, H. Blackwell, G. Jordan, R. Herring, and G. Lunnun. Mr. Henry Taylor, a London trades unionist, who had early evinced an interest in the rural bestirment, was elected secretary, and Mr. J. E. Matthew Vincent, the editor of the "Leamington Chronicle," who had also generously identified himself with the movement, and rendered it signal service by his journal, was appointed treasurer. In addition to this working organisation, the labourers very wisely accepted the co-operation of a number of gentlemen, whom they designated a consulting committee. Among the names on this committee appear some of the foremost patriots and philanthropists of

Britain, and their services have been at all times duly appreciated by the labourers, and have undoubtedly been of incalculable value to the movement.

The following code of rules was drawn up for the government of the young but energetic organisation, and accompanying each copy of the same was this address by the chairman :—

TO THE MEMBERS OF THE NATIONAL AGRICULTURAL LABOURERS' UNION.

In submitting to their brethren the Rules of the "National Agricultural Labourers' Union," the Members of the "National Executive Committee" have added certain Supplementary Rules, for the use of Districts and Branches. These Rules are not regarded by the National Executive as exhaustive, but simply as fundamental. It is felt that Districts and Branches should have perfect liberty to frame such laws for their own guidance as their own special circumstances may suggest ; that liberty is freely accorded, and the National Executive hope it will be exercised on the basis of the Rules for Districts and Branches, and in harmony with the General Rules of the National. The National Executive hope soon to see a Branch Union in every parish, and a District Union—that is, a combination of Branches—in every county or division, all communicating with a common centre, all observing the same principles, and all working for the same end. In the early stages of our movement, let the Branch and District Meetings be frequent, that enthusiasm may be kept alive, information be dispersed, and the Union be perfected. We must have no local jealousies, no self-seeking, no isolation. Unity of action is above all things necessary, and this can be secured only as all the Branches and Districts work through a common Representative and Executive Committee. We must have money, and we must have it in one central fund, to which all shall contribute, and from which, in time of need, all shall in turn be aided. The strength of the great trade societies is in their central funds. If we have a balance here and another there, it will be simply impossible to support a number of men in any emergency that may arise. We must have a common treasury large enough, through the payments of all, to support the demands that may be made in the interest of all.

The funds of a Branch or District would soon be exhausted if a number of men were thrown upon them, but the National Fund—the fund of all—would be rich enough to meet any demands which the National Executive might entertain, and to support our Members through any crisis. Let it be clearly understood, then, that the Branch remits its funds to the District; that the Districts remit three-fourths of their receipts to the National; and that any Branch or District failing to do this has no claim whatever on the general resources of the Union. The fourth allowed to be retained by the Districts can be disbursed at the discretion of the District Committee in meeting current expenditure and in promoting the general objects of the Union. For the working expenses of Branches an Incidental Fund is recommended, which may easily be realised by a small payment from each Member. Our movement has begun well. Success is, under God, in our own hands. Let us cleave to and work for the Union. Let peace and moderation mark all our meetings, Let courtesy, fairness, and firmness characterise all our demands. Act cautiously and advisedly, that no act may have to be repented or repudiated. Do not strike unless all other means fail you. Try all other means; try them with firmness and patience;

try them in the enforcement of only just claims; and if they all fail, then strike, and, having observed Rule 10, strike with a will. Fraternise, Centralise! With brotherly feeling, with an united front, with every District welded into a great whole, with a common fund to which all shall pay, and on which all shall have the right to draw, the time will not be distant when every Agricultural Labourer will have— what few as yet have enjoyed—a fair day's pay for a fair day's work. Nine and a half hours, exclusive of meal-times, as a day's work, and 16s. as a week's pay, are not extravagant demands. Society supports you in making them, and they will be met soon. Brothers, be united, and you will be strong; be temperate, and you will be respected; realise a central capital, and you will be able to act with firmness and independence. Many eyes are upon you; many tongues are ready to reproach you; your opponents say that your extra leisure will be passed in the public-house, and your extra pay be spent in beer. Show that their slander is untrue! Be united, be sober, and you will soon be free!

    (Signed) JOSEPH ARCH,
    Chairman of the National Executive Committee.

RULES AND CONSTITUTION.

Name.

1. The National Agricultural Labourers' Union.

Object.

2. (A) To improve the general condition of Agricultural Labourers in the United Kingdom.

(B) To encourage the formation of Branch and District Unions.

(C) To promote co-operation and communication between Unions already in existence.

Council.

3. A Council, consisting of one Delegate from each District Union shall meet at Leamington or elsewhere, as may be determined by the preceding Council, on the third Tuesday of May in each year, for the following purposes :—

(A) To elect an Executive Committee, together with a Treasurer, Secretary, and four Trustees.

(B) To receive a Financial Statement, with a Balance Sheet for the previous year, duly audited by a public accountant.

(C) To consider the General Report to be submitted by Affiliated Districts for the year ending 31st March preceding.

(D) To confer and decide on the general business and interests of the Union.

National Executive Committee—Composition and Functions.

4. The National Executive Committee shall consist of a Chairman, who shall have a second or casting vote, and of twelve Agricultural Labourers, seven of whom shall form a quorum.

5. The National Executive Committee shall seek the counsel and co-operation of gentlemen favourable to the principles of the Union, and shall invite them to attend its meetings, without power to vote.

6. The National Executive Committee shall meet each alternate Monday, and oftener if necessary—all meetings to be convened by the Secretary.

7. The National Executive Committee shall be entrusted with the expenditure of all moneys contributed by the public, by the Affiliated Districts, or otherwise; and shall employ the same in furthering the objects specified in Rule 2; it shall also adopt such general means as it may think

desirable to carry on the work of the Union, and shall appoint paid agents and officers at discretion.

8. The National Executive Committee shall make the necessary arrangements for each Annual Council, and shall submit a programme of the business to be considered to the Secretary of each Affiliated District, at least fourteen days before the third Tuesday in May. Should any important and unforeseen circumstances arise to necessitate such a course, the National Executive Committee may at any time convoke a Special Council, upon giving the usual notice, and shall do so forthwith on the written request of six District Committees.

9. The National Executive Committee shall communicate to the Secretary of each District Union any proposals or suggestions that may seem advisable in the general interests of the Union as a whole.

Settlement of Disputes.

10. All cases of dispute between the Members of the National Agricultural Union and their Employers must be laid before the Branch Committee to which such Members may belong; and, should the Branch Committee be unable to arrange the question to the mutual satisfaction of the parties interested, in conjunction with the District Committee, recourse shall be had to arbitration. Should the District Committee be unable to arrange for such arbitration, an appeal shall be made to the National Executive Committee for its decision. Any award made by arbitration or by decision of the National Executive shall be binding upon all members of the Union; and in no case shall a strike be resorted to until the above means have been tried and failed.

Financial.

11. The Funds of the National Agricultural Labourers' Union shall be invested in the names of the following

gentlemen as Trustees :—Mr. A. ARNOLD, Hampton-in-Arden ; Mr. JESSE COLLINGS, Birmingham ; Mr. E. JENKINS, London ; Mr. W. G. WARD, Periston Towers, Ross.

12. The Treasurer shall make no disbursements except on receipt of a resolution of the National Executive Committee, signed by the Chairman and Secretary ; at the first Meeting in each month he shall present a Cash Statement to the National Executive, and shall deposit, at interest, in the names of the Trustees, with Lloyd's Bank at Leamington, any sum in his hands exceeding £50.

District Committees.

13. District Committees shall bear the name of the County or Division embracing them; as, the "Kent District," or the "West Berks District of the National Agricultural Labourers' Union." District Committees shall be composed of Delegates from the various Branches of the District ; and each District Committee shall elect an Executive of seven Members, together with a Chairman, Secretary, and Treasurer, who shall meet monthly, and oftener when necessary.

14. Each District Committee shall regulate its own affairs in conformity with the general principles laid down in the preceding Rules ; but no Rules drawn up by any Branch shall be accepted by the National Agricultural Labourers' Union, unless they shall first have been ratified by the District Committee to which such Branch belongs.

15. District Committees shall do their utmost to prevent men who may migrate to another locality from underbidding their fellow Labourers already at work there.

16. Each District Committee shall be required at its own cost to send a Delegate to the Meetings of the Council.

17. Each District Committee shall send to the National Executive Committee, on or before the fourteenth of every

month, a brief Report of its proceedings; and on the third Friday in April, July, October, and January, under a penalty of 10s. for neglect, a Financial Statement, with the balance due on the quarter.

18. Each District Committee must inform the National Executive Committee of any important action contemplated within its jurisdiction; and should any proceedings be taken by a District without the sanction of the National Executive Committee, and be persisted in after the National Executive has signified its disapproval, such District shall not be assisted in its action by the funds of the National Agricultural Labourers' Union.

19. All Districts wishing to be affiliated with the National Agricultural Labourers' Union must remit three-fourths of the entrance fees and of the weekly contributions to the National Executive Committee, to be invested and employed in accordance with previous rules.

20. Cards of Membership, bearing the device of the National Agricultural Labourers' Union, shall be issued to the District Committees, to be supplied by them to their several Branches.

Branches.

21. Each Branch shall bear the name of the village or parish in which its business is transacted.

22. Branches shall consist of the Agricultural Labourers of one or more parishes in the same locality, who shall pay an entrance fee of 6d. and a weekly contribution of 2d.

23. Branches, as soon as practicable, shall unite in forming themselves into Districts.

24. Each Branch shall annually elect a Chairman, Treasurer, Secretary, and a Committee of seven Members, for the management of its business, to communicate with the District Executive, and, through it, with the National Executive.

25. Branches shall meet fortnightly for the payment of contributions and other business.

26. The Chairman shall preside at each Meeting of the Branch; he shall preserve order, promote the interests and repute of the Union to the best of his power, and sign all reports, minutes, etc. The Secretary shall keep the accounts of the Branch, record the minutes of all the Meetings, and pay over all funds to the Treasurer without loss of time. The Treasurer shall receive all moneys, and, under a penalty of 2s. for neglect, shall, in conjunction with the Secretary, remit them to the District Executive on the first Thursday in every month, together with an audited account.

27. Each Branch shall raise an incidental fund to meet its own working expenses.

28. Branches shall be at liberty to frame any bye-laws they may think necessary—regard being had to Rule 15.

29. These Rules shall be subject to additions or alteration only by the Annual or a Special Council. One month's notice of any intended Amendment must be given to the Secretary in writing, and such Amendment shall not be adopted unless two-thirds of the Delegates present approve it by vote.

The Union was now fairly launched, and, for weal or woe, a new force was henceforth to be brought to bear upon English society. Scarcely a month had elapsed since its consolidation at the May conference before a new and most potent agency was set in motion for the furtherance of its ends. On June 6th the "Labourers' Chronicle" was started by Mr. J. E. M. Vincent, and the circulation rapidly extended throughout the rural districts. In its columns week by week a series of very powerfully written articles appeared, and some of them — notably those from the pen of Mr. W. G. Ward, of Ross, on the land question—were of a nature to arrest the attention of other than agricultural labourers or their sympathisers. The influence of this journal upon the movement, and upon the great mass of the labourers generally throughout the country, can scarcely be exaggerated. It is no uncommon thing to see half a dozen labourers sitting under a hedge at their mid-day meal listening to a seventh who is reading from its pages.

INCIDENTS OF THE MOVEMENT.

In order to convey anything like a correct idea of this social revolt, it will be necessary to select from the ample reports of the labourers' organ some of its most striking incidents. As the work went on the opposition to it developed and intensified. How the work grew, and how this opposition was shown, will appear from a statement of what came under my immediate notice. Towards the close of the year 1872, as I was seated in my dining-room one evening, about half a dozen rough-clad labourers called to see me. On inquiring what was their business, they informed me that they had called to ask me to help them to form themselves into a Union. I advised them to well consider the matter, to "count the cost," and call on me in a week's time. At the end of the week they called again with a few more would-be unionists. Their minds were fully made up—they must have their Union. I drew up some rules for them and gave them a start, and as no school-room or public building could be on any account granted for their use as a meeting-place, I used to meet them once a fortnight in my coach-house. Things went on pretty smoothly for a few weeks, when one Friday evening a knock came at my front door, and on the servant going to see who was there she found a congregation of labourers outside. Could they "see the maister"? Certainly; walk in. "Here's a hallful of men, sir, wanting to see you," said the servant as she opened my parlour door. "Well, my men," I said cheerily, on going out to them, "what's your business to-night?" "Why, sir," replied one of them, "we're in a bit of a fix. The 'maister' has given us notice to-night that we must either leave the 'Union' or leave his farm by this day week, and we are come to ask you what we had best do in the matter." I paused before replying. It was no light responsibility to advise under such circumstances. Most of those men had families depending on them. Winter was close upon them; the employer who had thus come down upon them had taken good care to get his harvest well housed before he struck the blow; and leaving the farm meant leaving the home, and starvation. I told them to well look the affair over, pointing out the difficulties of fidelity to their convictions, at the same time urging them as much as I dared to "stand like the brave," assuring them that if they stood by the Union, the Union would certainly stand by them. In a week's time they came again. "Well," I said, "what's your verdict?" "To stand true to the Union, sir." "I am glad to hear it," I replied: "and now those of you who like I'll send to Leamington, and the secretary there will direct you to some work."

Immediately on their leaving the week previously, I had written off two letters—one to the secretary of the Union at Leamington, to be prepared for the worst; and another to the proprietor of the farm where the men were at work—the Earl of Radnor—to try and avert the blow. His lordship had, however, declined to interpose between his unwise tenant and his men.

Within a week from that time, most, if not all, the men were at work in the North at greatly increased wages; though unfortunately, through the lack of suitable accommodation—which has ever proved the weak point of migration—they had to leave their families behind them.

A few days after their departure, I had to listen to a torrent of abuse from the employer, who had found out, when it was too late, how egregiously he had blundered in his anti-Union crusade.

The next incident to which I can speak from

personal observation was one which involved a most important question of civil liberty.

Early in the year 1873 a labourers' meeting was held in the village of Littleworth, near Faringdon, Berkshire. At the instance of an ill-conditioned farmer of the village, three of the leaders of the meeting were summoned before the Faringdon bench of magistrates for obstruction of the highway. The case was exceedingly flimsy, as the Primitive Methodists had for many years been accustomed undisturbedly to hold their meetings on the selfsame spot. The men were, however, convicted. On my reporting the circumstances of the case to the executive committee at Leamington, it was unanimously resolved to hold an open-air test-meeting in the town of Faringdon. Mr. Arch and one or two gentlemen of the consulting committee determined to be present.

Towards the end of March this meeting was held. An immense throng of labourers half filled the spacious market-place of the quiet old country town. Mr. J. C. Cox, a justice of the peace for Derbyshire, who has distinguished himself by his chivalrous devotion to the labourers' cause from the first, took the chair; and by his side, in the waggon which constituted the platform, were the worthy president and secretary of the Union, Messrs. Arch and Taylor; Mr. W. Mackenzie, a London barrister; and two or three local friends of the movement.

Soon after the commencement of the proceedings a request was made to the chairman by the superintendent of police to dissolve the meeting, as it was illegal. Mr. Cox politely informed the official that the meeting was held expressly to test the illegality, and that it would not be discontinued till the proceedings were concluded; at the same time furnishing him with the names and addresses of all who were about to take part in it.

Contrary to the expectations of those whose charity was in excess of their acquaintance with human nature, especially that part of human nature which is represented by country squiredom, summonses were in due time served on each of these gentlemen "to be and appear" before the dread tribunal of the Faringdon bench "on the 15th April, 1873, to answer to the charge of wilfully obstructing her Majesty's highway on the," &c., &c.

Appreciating the importance of the point in dispute, the committee wisely determined to spare no expense in providing the defence. The eminent Queen's counsel Mr. Fitzjames Stephen, with Mr. Edward Jenkins, were retained for the occasion; and the case was entrusted to the able care of Messrs. Shaen and Roscoe, of London.

On the morning of the trial the usual quiet of the sleepy old town was disturbed in a most unwonted manner. Troops of labourers poured in from the villages. Everybody seemed to understand instinctively the issues of the battle. The future welfare of the Union was more or less involved in the struggle. With every public building closed against them, where could their meetings be henceforth held if their right of meeting on village greens and public market-places was successfully disputed?

So the plough and harrows were forsaken for the nonce, and Farmer Jones must wait till to-morrow for his barley-sowing, for Hodge is off to the trial.

And there were not wanting more distinguished visitors that day. The veteran reformer Charles Neate, Esq., senior fellow of Oriel college, Oxford, and a few years ago M.P. for the university city, was there. Editors of local journals and representatives of more ambitious prints were attracted by the bold venture of the squires. Birmingham had its reporter there, and the "wire" would bear to the great city itself a report of the important transactions.

"There was a petition I think you said, Mr. Clayden?" asked the astute Q.C., as he sat opposite to me at lunch on the morning of the trial, at the same time helping himself to a glass of Bass as unconcernedly as possible. "Yes," I replied, without the slightest idea of the significance of the reply; "I wrote it out myself in this very room, and delivered it to the chairman at the meeting, and heard it read." "That you have no objection to state in court?" "Not the least." "Thank you. A glass of sherry with you, Mr. Neate." Thus coolly did the distinguished lawyer settle the great point of his defence. In yonder court the leading farmers were congregating to hear their great Arch-enemy discomfited; for the word has gone forth that there will sure to be a conviction. The place of meeting was on her Majesty's highway—there could be no question of that; and there could be no question either that the concourse of people caused an "obstruction" of such "highway." Yes, the thing was clear as noonday, and they knew they could rely on the bench. Happy illusion! Within the capacious brain of the seemingly careless man before me lay concealed the little bombshell that would in a few hours' time blow all their little castles in the air clean out of sight. But I am anticipating my narrative.

At one o'clock on that Tuesday morning the little court-room of the Faringdon bench was packed as it had rarely been before. On the bench were the noble chairman, Viscount Barrington, the owner of yonder exquisite piece of Elizabethan architecture known as Beckett House. By his side is the sightless but indefatigable magistrate, Thomas Leinster Goodlake, Esq., one of the bitterest opponents of the Union that England could produce, and one whom it were better never to have been born than to offend. On the other side was William Dundas, Esq., a retired Chancery lawyer, whose judicious avoidance of any conflict with the energetic master of Kitemore makes him often appear to sanction what, as a truly kind-hearted and Christian gentleman, he must seriously disapprove of.

The fourth magistrate who is on duty to-day is William Campbell, Esq., the son of a wealthy New Zealand merchant and landowner, who owns the old-fashioned but highly picturesque property known as Buscot Park. Another magistrate occupies a back seat on the raised part which constitutes that more or less appalling tribunal the " Bench," but, for some cause or other, he takes no part in the proceedings. His name is Daniel Bennett, Esq., lord of the manor of Great Faringdon. His residence is yonder old Elizabethan pile surrounded with venerable elms. The broad acres which surround his domain, and which at one time formed a part of it, have, through the exigencies of time or the vagaries of its wanton whirligig, passed into the hands of the wealthy coroneted ex-banker who lives in the somewhat gloomy retirement of Lockinge, in the vicinage of the unromantic town of Wantage.

The other two gentlemen standing on the sacred precincts are Captain—or Colonel rather, I should say—Blandy, the chief of the Berkshire constabulary; and the vicar of the parish, the Rev. Henry Barne, M.A.

Seated immediately below the chairman is the bland and gentlemanly clerk of the bench, Mr. George Frederick Crowdy; and near him his equally polite political and professional rival, Mr. George James Haines, the local factotum of the Tory M.P., Col. Loyd Lindsay, who enjoys the reputation of being prospectively the richest commoner in England, through a happy alliance with the only child of the immensely rich Lord Overstone.

By the side of this deeply interested legal adviser of local well-to-doism sits the Reading lawyer to whom the conduct of the prosecution has been committed. He is a smart-looking middle-aged man, who may be fairly trusted to make the best of even a bad case, and if we may judge from his self-satisfied and somewhat jaunty air, he deems his case to-day anything but a bad one. On the opposite side of the table are seated Mr. Edward Jenkins, Mr. Fitzjames Stephen, and the celebrated London lawyer, Mr. Shaen. A quiet smile sits on the face of the former gentleman. Is he shaping out materials from this trial for another "Ginx's Baby," or "Little Hodge," or "Lord Bantam"? Very probably. He has got the *dramatis personæ* before him, if he did but know it, for a far more thrilling story than he has yet put in print. His chief, however, has no embryo romance before his mind. He is all business. A shrewd, wide-awake man that! Those full, heavy features, with that greater fulness above, are indicative of a reserve of intellectual force that will cut the comb of the Reading lawyer before the day is out, unless I am mistaken. Yes, gentlemen at the bar, your case is in good hands, or my penetrative powers are for ever humbled.

The case proceeds. The prosecuting counsel has stated his instructions, and his witnesses have made their record. A policeman or two avow there was an obstruction of the highway — a brewer of the town among others had informed them that he was obstructed in his passage from his residence to the brewery. He would be put in the witness-box to swear this. Later on it was found, however, that in a fit of discretion this valuable witness had left the constabulary in the lurch. Probably he had caught sight of Mr. Stephen's physique, or dreaded the powers of Mr. Jenkins' satire. Anyhow he did not appear—a fact of which Mr. T. L. Goodlake did not fail to remind the unhappy counsel for the prosecution in one of his most biting side-hits.

Another valuable witness, however—the landlord of the Crown Hotel—was not thus without the courage of his convictions. He was driving home, and had to rein in his horse before he could pass into his gates. The evidence of obstruction was conclusive! A third witness—a carriage maker of the town—was driving home, and although he could not say that he was obstructed, still, if he had wanted to drive into the Crown Hotel, he certainly should have had some trouble. Mr. Stephen did not deem it altogether essential that this valuable witness should be cross-examined. This must have struck the defendants as being rather careless. "That is my case, my lord," exclaimed the somewhat chopfallen lawyer, and resumed his seat.

As he did so, Mr. Fitzjames Stephen rose. It was

perfectly plain to the most superficial observer that the Q.C. had the game in his hand. But his trump card was not going to be played just yet. He must afford himself and the court the luxury of making mincemeat of the trumped-up affair before he gave it its *coup-de-grâce*. One witness—the burly superintendent of the police, facetiously designated by Mr. Stephen "the Village Hercules"—had referred to the fact that it was a service night at the church when the meeting was held, and godly parishioners had been prevented thereby from going to church.

Amid roars of laughter, in which the worthy vicar joined most heartily, Mr. Stephen appealed to the reverend gentleman before him whether the hindrances to worship were not probably more internal than external. Of course the mysterious non-appearance of the chief witness—the brewer—did not escape the satire of the learned gentleman, nor the valuable contribution to the case from the homeward-bound coachmaker. Long before the speech was ended, there were not six men in the court who did not feel confident that the prosecution had collapsed. And when, with that ineffable satisfaction with which an absurdity is finally disposed of, he quietly informed the Bench that even if an obstruction was proved the prosecution was powerless against his clients, as they were protected by the Bill of Rights, it seemed a superfluity to put witnesses in the box. However, Mr. Charles Neate very clearly disposed of the rubbishy testimony as to an obstruction, and I was able to settle the question of a petition, and on the completion of the defence, a very short period of retirement from the court was sufficient to enable the Bench to declare the case dismissed. Though inevitable, the victory, however, was not the less valuable. Henceforth the right of public meeting in towns and villages was secured to the Union, on the simple condition of having a petition to Parliament on the programme. The Bill of Rights protects from imprisonment all assemblies of English citizens who meet for the purpose of petitioning the House of Commons.

Returning for a moment to the year 1872—the natal year of the "Union"—there were two or three incidents which deserve notice. On the 2nd September the Bishop of Gloucester—Dr. Ellicott—made a very indiscreet reference to the movement at a meeting of the Gloucester Agricultural Society. Severely animadverting on the "agitators"—as he designated the delegates who were deputed by the Leamington committee to proclaim the new social gospel throughout the rural districts—he significantly suggested the village horse-pond as a not unfitting place in which to plunge the wretched disturbers of the peace. Never was *lapsus linguæ* more unfortunate; and still more unfortunate was the dogged withholding of apology by the bishop. From a thousand platforms and in a thousand publications has this fatal utterance been repeated. It has travelled all over

Canada, and is as familiar to the Americans as any saying of our Gladstone or Disraeli. Like many another offender, the unfortunate cleric has again and again pleaded that it was "only a little one." But the plea has been useless, and doubtless, small as the offence seems when viewed as the semi-jocose post-prandial utterance of a somewhat out-of-place clergyman, it has probably done as much as anything to give the movement its present decidedly anti-clerical complexion.

By the first week in September the subscriptions in aid of the Union, as published in the "Labourers' Chronicle," had reached nearly a thousand pounds. A considerable number of Unionists emigrated towards the end of the month to New Zealand, under the auspices of the eminent railway contractors Messrs. John Brogden and Sons, whose terms were of an unprecedentedly favourable character, owing to their demand for men to carry out their railway contracts in that colony.

On the 12th of November an electoral reform conference was held in St. James's Hall, London, under the presidency of Mr. Joseph Chamberlain, of Birmingham, when the question of the enfranchisement of the agricultural labourers was ably discussed. Mr. Arch was present, and replied with his wonted force to the objections raised on the ground of the illiteracy and general unfitness of the labourers for the political boon.

A month later a great meeting was held in Exeter

Hall to express the sympathy of the London citizens in the movement of the rural districts. The Lord Mayor—Sir Sydney Waterlow—was to have taken the chair, but, owing to a discovery that Messrs. Bradlaugh, Odger, and others of the extreme school of politics were intending to seize the occasion for the airing of their peculiar views on the land question, he was persuaded to decline the honour. Happily, Mr. Samuel Morley, who had just previously given the munificent sum of five hundred pounds to the Union funds, at once filled up the gap, and thus, in a twofold manner, testified his respect for a movement which was at that time far from being universally popular. The chief speakers on the important occasion were Messrs. Arch, Ball, and Mitchell, as representing the labourers; and Dr. Manning, Mr. Thomas Hughes, M.P., Sir Charles Dilke, M.P., Sir Charles Trevelyan, Sir John Bennett, and Mr. Mundella, M.P.

The admirably-expressed story of the labourers elicited the utmost sympathy, and but for an unfortunate and most ill-timed invasion of the platform by a pushing secularist lecturer, the demonstration would have proved a complete success. A resolution was passed amid considerable enthusiasm, declaring the present condition of the agricultural labourers a national disgrace and inimical to the best interests of the nation, and in favour of speedy measures for their social and intellectual amelioration. On the motion of Mr. Charles Bradlaugh, a rider was added in favour of certain changes in the land laws.

The progress of the Union during the year had been of the most remarkable character. From village to village, and from county to county, it had spread with a rapidity wholly unprecedented. The delegates—for the most part labouring men, and to a large extent men who had gained their power of speech through their connection with the village chapels of the Primitive Methodists—threw into their work the intensest enthusiasm. Mr. Arch, whose services were of universal requisition, had performed simply Herculean tasks, speaking to immense meetings night after night in widely-separated districts. The result was a universal interest in the question, and thoughtful men were asking one another whereunto the thing would grow. Landlords and farmers united in heaping upon the unusual phenomenon in their midst their sarcasm and abuse, and the whole tribe of petty journalists in country places who live upon the smiles of the wealthy and well-to-do classes joined in the chorus of contempt of the "rural upstarts."

In order to convey a correct idea of the kind of work done by these agents of the Union, and how they do it, I will give a few extracts from the "Labourers' Chronicle," which gives a weekly digest of the proceedings of the delegates.

Perhaps it may be as well to preface these extracts and quotations with a sort of manifesto which the

"Chronicle" issued the week after the first anniversary of the Union had been held in Leamington. A controversy had sprung out of a leading article of the "Chronicle," which will be found farther on, and the following straightforward deliverance was the result :—

The "Labourers' Union Chronicle" was a private venture, risking much with a problematical gain; commercial notions were not supreme in directing its birth. We had an idea, and we were, and are, willing to sacrifice much for the idea. We mean to stand by the weakest member of the British family, guide him by our counsel, educate him in true manliness, and be his champion against all assailants.

Our programme we first set out with we shall stand by to the end; and that we may not be misunderstood, often as we have stated it, we will state it again :—

1st. A rise in wages. We dictate no amount; we place no limit; we simply say that we think, looking at prices of produce and fuel, 16s. a week is not too much for every able-bodied farm labourer; shepherds and waggoners of the right sort cannot be overpaid at £1 a week. As to the representation of farmers that the moment a man joins the Union he is not a free agent, it is utterly false. If a body of Union men choose never to ask for an advance, they are never coerced; if they make a foolish bargain, they are neither fined nor condemned. We give them good advice upon all points as far as we can; when they want assistance, it is granted. If they strive lawfully and respectfully for an advance in wages, and are met unfriendly, the committee will stand by them, faithfully protecting them at any cost.

2nd. We desire and strive for the franchise for the labourer. If he is to respect himself, and be respected by others, he must have the dignity of citizenship; and as Parliament for five hundred years has oppressed him, it is quite time he should be on a political equality with the landowner, his old oppressor. Taxation without representation is tyranny—is as true from his mouth as from the middle-class man in the old unrepresented boroughs.

3rd. The farm labourer must have a fair chance of land ; it is required not only for his benefit, but for the benefit of the nation. The land monopoly must be destroyed, and the land nationalised ; we advocate its being done slowly and moderately, but we desire the change to commence at once. We advocate the extension of Mortmain Acts to the extreme limits ; in other words, all lands held by charities or corporations of every sort and character should be seized at once by the Government, five per cent. less than their net produce should be paid to the late holders, and five per cent. less each year until there is no payment at all. In the case of useful schools and colleges, the payment of the full net produce in perpetuity may be a useful national benefit.

Then all waste land should be gradually divided up in four, eight, ten, and twenty acres, as the soil and district and demand indicate as best ; houses to be built, and capital advanced just as has been done to the large landowners ; four per cent. rental in perpetuity to be paid upon land and permanent buildings, five or six per cent. upon money repayable to cover capital and interest in twenty years.

All Government forest land to be used up in the same way as fast as there is a demand.

And the most important point is an extension of the probate duties to land. Every person leaving by will, or giving by deed of gift, over one hundred acres of land to each person, the receiver shall have cut off one-tenth, to be nationalised and used as stated above.

This simple and moderate remedy would effectually destroy the land monopoly ; primogeniture and entail would vanish before it without any declaratory acts of Parliament. And the land would be gradually nationalised; in about two centuries no one would hold more than a hundred acres of land, and nearly one-half of the land would have been confiscated to Government, giving a revenue of twenty-five millions a year; and the benefit would commence at once, and be continually increasing as fast as such changes can healthily proceed.

SPEECHES OF UNION DELEGATES.

The following is the substance of an address lately delivered at Nutfield, Essex, by Mr. Heywood, one of the delegates of the National Union. He said that in the remarks he was about to offer he should address himself exclusively to the working men present, and he should therefore, perhaps, be rather plain, or as St. Paul had said, "rude of speech." He was very pleased to see such a good meeting and to meet with them on this important question. He was especially glad to see so many women present, as he found that they were everywhere great friends to the Union movement. He had known one woman who threatened that if her husband didn't join the Union she would tear all the hair off his head. He supposed none of those present wanted bald heads. Then they had better join the Union. The movement, since its inauguration, had made rapid progress. Thousands had been gathered into it, and the Union was daily increasing in strength. They were not, however, getting on so well without opposition. The farmers were, in most places, violently opposed to them. At a meeting held at Fletching the farmers did not know how to behave themselves like gentlemen. For such conduct as that there was no excuse. He (the speaker) always endeavoured to say nothing but the truth, and to say it in a becoming manner. The farmer who opposed a man for becoming a Unionist was disloyal to the Queen, for the Union was according to law. The working men did not like to be called disloyal to Queen Victoria, though a farmer could not object to the Union without committing an act of disloyalty to his sovereign. His (the speaker's) advice to the labourers was, when they found themselves opposed by their masters, not to fly to the Law of Moses, "an eye for an eye and a tooth for a tooth," but rather to the law of the gentler dispensation of the Gospel; and not to retaliate, but to bear the enmity against them like men, and, if possible, like Christian men. There was no strength in swearing or in bad language, but in truth and law-abiding manliness. The labourers should be respectful to their masters, whatever they may be to the labourers. The speaker, continuing, said that wherever the Union had been established wages had gone up; yet there were men who were not in the Union who could still hold out their hand for the extra shilling on Friday night. It was the Union that had got that for them, and it was not honest of them to receive the benefit and refuse to join. Moreover, they must remember that the higher wages were not only to be got but to be kept, and it was only by means of the Union that that could be done. His advice, therefore, to all the labourers was to join the Union at once. If they didn't they must mind the hair on their heads. The wives wanted the extra pay, and girls were not to blame if they refused to marry non-Unionist men. What he wanted for them was, not only to be better paid, but to be better men; to have not only more to eat and drink, but to be better educated and so better clothed, and to live in better cottages. The Union, therefore was likely to give an impetus to work, for the better cottages would demand the services of the carpenters and the plasterers, and the masons and the bricklayers.

There was another thing he would remind them of, and that was, not to neglect the franchise, nor to rest till they had obtained it, as it was that by which they would be enabled to raise themselves up in the political and the social scale. They must not neglect the claims of the mind in favour of the claims of the body. They must learn to read, read the Bible, read good books, and good newspapers; and they must not only read, but read thoughtfully; not be like the butterfly that flies from flower to flower, but like the bee that delves down laboriously into the honey-cup and sucks its treasures from the very bottom. The Unionist men had been accused of being republicans and infidels, and Sir Michael Beach had written to the "Times" to caution the public to distinguish between the non-Unionist labourers and the Union delegates. He repudiated the assertions of Sir Michael Beach as unjust. They were not republicans, but loyal subjects of the Queen; they were not infidels, but believed in the great truths of Christianity. He, for one, might tell Sir Michael Beach that he (the speaker) had preached the Gospel, and that he loved his Bible. He would advise their oppressors to study that book more. Let them take Cruden's Concordance, and turn to the word "oppression," and let them refer to their Bibles and see what was said about it there. Let them search through the whole of the Old Testament and the New, and they would find that though much was said against it nothing was said for it. And what had been God's way for the deliverance of the oppressed? When the Israelites were to be brought out of their cruel bondage in Egypt, had He not placed His hand upon Moses, the son of one of the oppressed people? When God Himself walked this world to bring mankind into the liberty of the Gospel, where did He go? Did He go to Athens? there were poets and philosophers there. Did He go to Rome? there were emperors and warriors and statesmen there. No; but He went to the poor fishermen of Galilee. And when the time came that the American slaves were to be liberated from thraldrom, did He not place His hand upon Abraham Lincoln, the man who in early life had followed the humble occupation of a rail-splitter? When, too, the agricultural labourers were to be delivered from their long oppression, it was not to be done by our great men, our philosophers and generals and statesmen; but God had put His hand upon Joseph Arch, the Warwickshire labourer. He (the speaker) would urge upon all present to join the Union, which he believed was calculated to better their condition. Its principles were good, its objects were good, and there was everything to encourage it. It would improve them, would better their conduct, and protect them. It was said that the mechanics were more intelligent and better educated than the labourers: this was to be attributed to the principles of Unionism; and he hoped the mechanics would continue to help the labourers in this movement as they had done hitherto. Let the shopkeepers, too, with whom the labourers spent their money, encourage the Union men. They would equally reap the benefit of the higher wages. And let them try, too, to get the co-operation of some of the wealthy and influential gentlemen of the district. Lastly, when at length they had reaped the advantages of the movement and began to feel their power, he would say to them, "Never be tyrants." Let there be too much of moral dignity about them to retaliate. A farmer in Berkshire had said that what they (the farmers) wanted were "men with strong arms and no brains." That was an insane expression. He would like to put that farmer into a railway train, shut him up by himself, lock the doors, and turn on the steam. He would perhaps want to go to Manchester or to Liverpool. "Well," he (the speaker) would say to him, "go a-head; the steam's up, there's no driver, and you can't get out, but you've a strong arm and no brains; away you go to Manchester, or Liverpool, or whatever place you want; you've a strong arm and no brains." He (Mr. Heywood) contended that farm work was skilled labour. Let the man who thought it wasn't try his hand at mowing, for instance. Why, there was skill even in stone-breaking, which was generally supposed to be the last thing a man took to before going into the workhouse.

At the conclusion of the speech, which was attentively listened to throughout, three ringing cheers were given for the National Union, for Mr. Arch, and for Mr. Heywood. A large number of labourers were enrolled as members of the Union at the termination of the proceedings.

REIGATE, SURREY.

On Monday evening, May 5th, Mr. Hayward (delegate) addressed a small meeting of about fifty persons in the Market Place. A drizzling rain was falling, and the meeting had not been widely advertised; otherwise, there is no doubt that a larger number would have been present. Mr. Hayward commenced his speech by saying that every class had its own grievance; it would not, however, be necessary to particularise the landowners' grievances, as they could always speak for themselves. The farmers had great grievances—the first of these was a bad system of land tenure. Many farmers held their land on a six-months' notice, and, therefore, if a man took a farm that had been badly cultivated, and expended a great deal of capital upon it, so as to make it more valuable, he might receive a six-months' notice, and could get no compensation. He would have to leave behind him in the ground all his capital. The natural consequence of this was that the land was under-cultivated, as the farmer could not be expected to invest his capital without security. This was a grievance not to the farmers only, but to the whole nation, as the land was rendered less productive than it might and ought to be. Then the Game Laws were a great grievance to the farmer, as he was liable to have all his crops eaten up by game, and could get no redress. The National Agricultural Labourers' Union was no enemy to the farmer, as it would tend to get their just grievances redressed, as well as the grievances of the labourers. The farm labourers had many grievances. In the first place, they were underpaid; and the first demand of the Union was that a fair day's pay should be given for a fair day's work. He did not profess to state the exact amount which the labourer should receive; it was impossible to do that, as the cost of living varied in different places; but the labourers ought to receive sufficient to provide for themselves the necessaries of life, for, because the labourer was underpaid, he was consequently underfed. He himself had frequently gone to work in the morning with nothing but a piece of dry bread, which he had been obliged to soak in the water in the fields to make it go down; he had sat down to a Sunday dinner of nothing but potatoes, bread, and salt, without a scrap of meat. That was not sufficient food for a hard-working man. Then the Union demanded that the labourers should have proper cottages; it was not right or decent that a married man and his wife and their growing-up sons and daughters should all be obliged to sleep together in a single room. They demanded that cottages should be supplied with separate sleeping rooms for the two sexes. Then there was the question of the employment of women. The Union demanded that women should not be employed upon work that was only suited for men; they demanded that, as a rule, women should not be employed to work in the fields. At times, when there was a pressure of work, it was quite right that women should be employed rather than that the fruits of the earth should not be gathered in; but, as a rule, women should stay at home. He had seen women carting and spreading manure; that was not a woman's work. He had seen women work at the tops of machines, with their petticoats blown about by the wind; that was not a proper or a decent place for a woman. Lastly, the Union demanded that the agricultural labourers should have the franchise. Parliamentary leaders had to consider the opinions of the artizans in the great towns, such as Manchester and Birmingham; but they cared nothing for the opinions of the labourers, because they had no political power. This was neither right nor just. By means of the Union, the labourers were getting power and influence, and some of their wrongs were being slowly redressed. He exhorted them not to be tyrannical, but to use their power to gain what was just for their own class, and not to oppress other classes. All had their rights, and the rights of all should be respected.

The above is only a very brief outline of a most able and interesting address, which was listened to, with marked attention and interest, by an audience composed chiefly of labourers and artizans. At the conclusion of his address, Mr. Hayward asked if any one would like to address the meeting.

The Rev. J. P. Knight (incumbent of St. Luke's (Reigate), said he would like to say a few words. He had attended that meeting because he had always taken an interest in the National Agricultural Labourers' Union. He thought it had a right to the name "National," because it did not claim undue privileges for the labourers, but only that justice should be done them; and, apart from all questions of right or wrong, it would be a national advantage that justice should be done to the labourers.

CAPITAL MEETING AT BALDON.
ADMIRABLE SPEECH BY A CLERGYMAN.

Mr. C. Holloway (delegate) writes (April 22nd):—According to appointment I attended a meeting at Baldon last night, which I think, upon the whole, was one of the best Union meetings I have hitherto attended. Baldon is one of those few villages which have not suffered from the overreaching and covetousness which have for years been engaged in robbing the poor and helpless of the village green and common rights. Here, in the very heart of the village, is a beautiful and magnificent green, comprising within its limits, I should say, sixteen or eighteen acres of good pasture land, well studded with beautiful elms; springs of water bubbling forth and dancing gracefully in the glad sunlight, pursue their onward course right across the green, while all around, encircling this green, stand forth, in bold relief, the cottages of the villagers, each of whom has the privilege of using the green as a public recreation ground, and also the right to turn sheep, pigs, or geese (if they chance to have any) out to graze upon the greensward. May the day never come when these poor villagers will be robbed by their rich neighbours of this piece of common land, and may it for ever be preserved as a memento of what the English cottager used to enjoy in the good old times long since gone by. At the time appointed, in company with a few friends, I took my stand upon this green, and, sir, it did my heart good to see the people come flocking around me from all parts, and very soon I was surrounded by a very numerous and attentive assembly. Previous to opening the meeting I was informed that the clergyman of the village was present; I immediately invited him to take the chair, and preside over the meeting. This he very courteously declined, upon the grounds that he had come as a hearer, to hear for himself the complaints and grievances of the labouring classes as set forth by one of themselves, and then, having heard both sides of the question, he should be able to judge of the merits and justice of our cause. Mr. John Carter, a labouring man, was then voted to the chair, who, in a very able speech, alluded to the wrongs, privations, and sufferings, of the labouring classes, referring very touchingly to his own experience as to what he had endured during two long winters in which he was incarcerated in that modern prison—I mean the union workhouse at Abingdon—and the poverty and distress he has had to contend with nearly all his days, caused by the iron hand of tyranny and oppression. He then pointed triumphantly to the Labourers' Union as the only means of emancipation for the oppressed labourers of our country; after which he called upon Mr. Charles Stile, of Littlemore, who gave us a very neat and appropriate little speech. Mr. C. Holloway, the Union delegate, next addressed the meeting in his usual forcible style for about an hour and a quarter, frequently eliciting loud applause from the immense multitude. At the close of his speech he was told by the chairman that the clergyman wished to say a few words. The rev. gentleman was very cordially introduced to the meeting, and, after the cheering with which he was greeted had subsided, proceeded to make the following remarks. He said he had listened with very great pleasure and interest to the speeches that had been made, and to the arguments which had been brought forward by Mr. Holloway in support and defence of their great movement. He believed that the labourers had been badly paid; that in this age of progress the condition of the labourer had not kept pace with the times. He believed, from the statements made by Mr. Holloway, that the Union was designed to benefit and bless the people; and he could say "God speed' to any movement which had for its object " the glory of God and the bettering of the position of the masses of the people, both temporarily and spiritually." He said that he had listened very attentively to Mr. Holloway's lengthy speech, and although much had been said, yet not one word had been advanced which he should like to recall; and after noticing the different points which had been brought forward, he said that he should go home from that meeting more favourably impressed with the principles of unionism, as the only means whereby the working classes can be uplifted to their only true and proper position. He then dwelt at some length upon the advantages of migration and emigration; referred to the abolition of the Game Laws, and other improvements which might be brought about for the advantage of the community at large; and then brought his very interesting and practical speech to a close by inviting the men present to act upon the good advice which had been given by Mr. Holloway,—that is, said he, when you have got an advance of wages, don't take it to the public-house and squander it away, but take it home to your wives and families, and by so doing the cottage homes of England shall once more become the abode of the brave, the free, and the happy. This interesting meeting was then brought to a close by giving three hearty good cheers for the speakers, and three for Mr. Arch and the Union.

OXFORD DISTRICT.

Mr. C. Holloway, of Wootton, writes :—I once more take up my pen to address a few lines to you on the prospects and progress of Unionism in our district. After having, with Mr. Leggett, held a very successful meeting at Chalgrove, on Monday last, I walked over to Tetsworth, on Tuesday, where, in the evening, I held a meeting that was very orderly and very successful. T. Johnson, who is just appointed chairman of the newly-formed branch, ably presided over the meeting. I then addressed the numerous assembly for more than an hour upon the principles of Unionism, its constitution, and its objects. The people listened with the greatest attention, very frequently applauded, and when at the close of my address I invited the men who wished to be free to step forward and join the Union, sixty-two came to the front and enrolled their names, paid their moneys, received their cards and rules, formed a committee, and put this new branch into good working order. We then brought the meeting to a close by giving three hearty cheers for Mr. Arch and the Union.—On Wednesday, April 2nd, I walked from Tetsworth to Shillingford, for the purpose of sowing the seeds of Unionism in that village, where no agitator had been before. I met a large assembly of people, and pointed out to them the past position of the farm labourer, alluding to their poverty, their destitution, their privations, and their sufferings, with no prospect before them but the workhouse in which to end their days, and no place of rest in this world but a pauper's grave. I then pointed out to them the advantages of Unionism, showing that it was only through Union that the chains and fetters by which they had been bound down in serfdom could be snapped asunder, and the only means whereby they could be elevated to their proper position in society. I noticed in the middle of my speech some loud talking among four or five respectable-looking men. I afterwards learned that these were farmers, and that previous to the meeting they had planned a scheme to fill their pockets with rotten eggs, with which they meant to pelt the speaker, and so break up the meeting; but the plot was known to the labouring men, so they formed themselves into a circle around these farmers, and when one of them made a proposal to interrupt the speaker, one of the labouring men stepped forward, shook his fist in his face, and dared him to hurt one hair of my head. "Oh," said the farmer, "if that's it, we must be quiet." They stood till the meeting was nearly over, said the speaker spoke well, that what he said was true, and then quietly took their departure. After talking to the people for over an hour we closed the meeting with three good cheers for the Union, and with a promise that I should visit them again soon.— Thursday, April 3rd, I walked from Shillingford to Nettlebed; here we held a very large and enthusiastic meeting, some five hundred persons being present. It was the first meeting that had been held there, and although there were so many people present, yet I have never seen a more orderly, better-conducted assembly of men. An employer of labour here threatened his men that if any of them went to the meeting he would "sack" them, if it was the best man that he had. Yet, sir, notwithstanding this threat, some eight or nine of these men had the impudence to disobey their employer's orders by attending the meeting; and I told them not to mind their employer's threats, but assert their rights, their freedom in a free country; and if their employer "sacked" them if they joined the Union, "the Union would back them, and off to the north it soon would pack them." Another incident took place here, which I think is without a parallel. A married man with a wife and six children told a friend of mine that 9s. per week was plenty for any man; for his own part he was very well satisfied, and couldn't see how the farmers could afford to give more. Now, sir, allowing 2s. per cwt. of coals, and 1s. 6d. rent of cottage, that would leave the small pittance of 5s. 6d. for his wife to provide (allowing three meals a day for eight persons for seven days) 168 meals for eight persons for one week. If that wouldn't puzzle the most accomplished French cook to carry out, I want to know what would. And there would be nothing left for shoes, clothing, sick-clubs, education, and all the other little incidentals which we all know are so essential to make a home happy and comfortable. I told them that the fellow ought to have penal servitude for life, and the people all shouted out "Hear, hear." I talked to this large assembly for an hour and a half, fully explaining the principles of Unionism, and exhorted them all to join the Union. I told them that we had been serfs long enough; that we had been slaves long enough. Oh! my heart revolts and my spirit sickens when I think of the indignities which have been heaped upon us as a class. Oh! to cast off this serfdom, to break these fetters, and to stand forth in all the dignity of true manhood, socially, morally, and intellectually, as God hath willed it! I can conceive no honour greater, no joy more complete. At the close of the meeting several joined the Union; and I think that Nettlebed will form a centre for a good branch.

GREAT DEMONSTRATION IN NOTTINGHAM MARKET PLACE.

A large and thoroughly earnest audience of Nottingham mechanics and artizans, to the number of about four thousand, assembled on Tuesday evening in the fine old Market Place, to listen to addresses from Mr. J. C. Cox, and Mr. H. Taylor, claiming their sympathy for the agricultural labourers. Messrs. Arch and Cox had addressed a large meeting in this town about two months previously, and they met with so much sympathy that it was determined to again appeal to the inhabitants now that the Union was in a critical condition. Nor has their confidence been misplaced; several pounds' worth of coppers were collected at the meeting. The warmest feeling of fraternal sympathy was shown on all sides, and we may shortly expect further practical proofs of it from the collections that are being made by the different trades societies of the town. The meeting was announced to commence at 7.30, but the yeomanry band was in possession of the ground at that hour. At eight o'clock, however, they retired to aid by their music the appetites of their officers at mess assembled, and left the field clear. The speeches were all well received and listened to with much attention. The special point of the meeting seemed to be the landlords' prayer from the authorised primer of Edward VI., which evidently much astonished the audience. Mr. Cox read it in a loud and reverent tone, and the people at the conclusion joined in one of the loudest and heartiest "amens" that it has ever been our lot to hear. Mr. Cox was loudly applauded when he expressed his wish that all the State clergy were now obliged to read that prayer Sunday by Sunday. At the conclusion of the proceedings a deputation waited on the Nottingham Trades Council to explain the present position of the National Agricultural Labourers' Union, and met with a most brotherly reception.

In opening the proceedings, the Chairman (Mr. Councillor Egginton) said they had assembled on the present occasion to express their sympathy with those who were deserving of their help—viz., the agricultural labourers; who, as a class, had been for many years the worst paid and the most downtrodden. In doing so they were not met to say a word, so far as could be avoided, against any other class. They did not believe in causing the separation and division of classes, thus setting one against another; but they did believe that a vast amount of good might be done by the promotion of a better feeling. As long as differences and jealousy existed, so long must they expect to see combinations. If the tenant farmer could only see his own interests properly, he would rather join in with the movement, because it was only those that had taken an active part who knew the difficulties with which the tenant farmers themselves had to contend. But gentlemen would address them who could say more on the subject than those who, like himself, were living in towns. It was their desire to elevate those who were toiling day after day, and year after year, in almost hopeless misery. After expressing his cordial sympathy with the object under consideration, he would not, however, now detain them, but would leave it to others whom they would no doubt be glad to hear.

Mr. Coburn, who followed, spoke of the pleasure which it gave him to take part on the present occasion. After some other preliminary remarks, he moved, "That this meeting, deeply sympathising with the agricultural labourers in their endeavours to elevate themselves from the low condition in which they have been for centuries, and believing union to be the best means of accomplishing that object, pledges itself to support them to its utmost." The resolution was one which must commend itself to their attention. He thought the miserable condition of the agricultural labourer was a scandal to English Protestantism and civilisation. Nothing scarcely could be imagined more shocking than the rate of remuneration which they received for their labour —in some places, at all events, where they were paid 9s. or 10s. or 11s. per week each, to maintain themselves, their wives, and children, and to provide clothing and a nominal education. He would ask such as had only been in the suburban villages to form an idea of what the downtrodden labourer was—villages, too, where there might be opportunities, which could not be had in the remote rural districts, for ameliorating their position. What did they generally see in an English village? A poor, hulking, miserable fellow, who was half frightened to death if he saw a gamekeeper, or a policeman, or the squire, or the parson. Was such a one a sample of an Englishman they would like to point to? If not, whose fault was it? It was that of the system, and partly that of themselves, for allowing a man to vegetate, as it were, under such a system. But to look at the reverse case, let them see what the man was made by the recruiting sergeant. It might be said that the remedy was to be found in emigration. But to him nothing was more distasteful. If England could not support its sons, let them cease to regard it as they did; let them cease to pride themselves on its fame and glory; let them obliterate those records of enterprise and successful efforts which had spread throughout the world. He deplored the state to which the agricultural labourers had been reduced. Mr. Joseph Arch, on his visit to the town, compared their treatment with that received by animals —he compared their accommodation with that provided by the country squire for his horse or dog. Often animals were fed and bloated out of their own proper proportions to gratify some man who wanted to take a prize for a brute about fifteen times uglier than God ever intended it to be. He wondered Englishmen did not rise in their strength and give one vigorous blow from the shoulder at the system to which he referred. The speaker, among some other observations, alluded to the fact that the ex-Sheriff, Mr. Booth, had forwarded to the Union ten guineas for their object; and at the conclusion of an effective address he was cordially applauded.

Mr. J. C. Cox, who met with a very hearty reception, then said that he had great pleasure in again appearing before the Nottingham audience on this question, which he considered to be one of vital importance, not to the well-being of any particular class, but to that of the whole nation. They might wonder, after so recent a visit from Mr. Arch and himself, why they should once more trouble the good town of Nottingham; but their former reception was so kind, that they came to try if it would again be the same. The Union, at present, was to a certain extent in a position which required that they should come before the public, to endeavour to be on a par with the great agriculturists of the country, who were seizing this opportunity, at a time when the seeds were pretty well in, and when there was an interval of five or six weeks before the hay harvest, to aim a blow at the Union. Most of the lock-outs during the last few weeks had been not owing to any extortionate demands for increased wages on the part of the men themselves, but owing to the farmers turning the men out because they had refused to sign papers to give up the Union. It was thought that now was their time of weakness. But those of his hearers who were trades unionists (and very many of them in Nottingham knew the benefits of such societies when properly managed) would be ready to extend the hand of sympathy to aid in opposing that of which he had spoken. They were not in the least desponding about the issue of the struggle. They believed that they should tide over the difficulty; and they might depend upon it that so soon as harvest time approached, if not before, the labourers would be able to get the advantage of their union. It was always gratifying to any one engaged, like himself, in such a movement, to find material results had accrued, and he rejoiced that these could be seen already. In Dorset, he believed that at present the wages were on an average, taking all round, about 1s. 9d. a week higher than at this time last year, when the Union had not obtained its hold. Though to persons, say, in his own position such an increase might sound exceedingly trifling, yet to those in the position of the agricultural labourers it was almost a case of life or death whether they got it. In the churchyard of a village in Dorset, a labouring man stood on his father's grave and assured him (the speaker) that his father had died because he could not make up his mind to go into the workhouse when in want of bread. Though a labourer might have worked the best years of his life, and lived with the utmost frugality, it was impossible for him to put by a single sixpence for provision in case of need; and, consequently, the moment his health began to fail, or old age came upon him, there was nothing left for him but to seek the workhouse or obtain poor-relief. Was not that a most awful, scandalous, and wicked state of things in our so-called Christian England? He agreed generally with the remarks which had been by the previous speaker. Turning to another matter, he found that Mr. Disraeli, in writing to a body of Conservative working men (if, indeed, there could be such persons), congratulated them on their adherence to the great and noble principles which had made England so prosperous and happy. That this was so he himself would deny. It was not a prosperous or happy country when an immense number of persons in both the town and country had to eke out a miserable existence on scanty wages. He did not hold with what were ordinarily understood to be Communistic or Socialistic views, but at the same time he must say that he thought it fearful when the distribution of wealth was so disproportionate as now in this country. He asserted, fearless of contradiction, and on the best authority, that at no time was the disproportion between the rich and the poor so great as now. It had been said that the rich were becoming richer, and the poor poorer. Certainly, in various directions there had been advances, to artisans and mechanics, but these were not in proportion to the increased cost of living and of clothing, with other necessaries. This was an important question, and one which deserved consideration. Next the speaker went on to advert, at some considerable length, to the land question, in reference to which he alluded to the matter of the inclosures of commonable lands, asking what might have been the result if a portion of these had been appropriated for the use and maintenance of the poor. A valuable lesson might be learnt from the history of the inclosures of commons. It had been estimated that there are in the United Kingdom 10,000,000 acres of uncultivated land, which would amply repay cultivation; and this he pointed out as being an important matter. After referring to the happy results which had followed in Suffolk, where a landlord had, under suitable conditions, let land to a number of his labourers, Mr. Cox touched upon the position taken up in reference to this movement by the clergy of the Church of England, among whom, however, there were noble exceptions, like that o Canon Girdlestone. Having read a prayer from the Prayer-book of the Church of England in the days of Edward VI., which had especial reference to landlords, the speaker urged eloquently the claims of the Union to public sympathy and support. He noticed the contribution made by Mr. Booth, and also one of £500 by Mr. Samuel Morley, to the Union, and ended by seconding the resolution. On resuming his seat he was loudly cheered.

Mr. Taylor, the secretary to the Labourers' Union, said he was happy to tell them that the Union had met with great sympathy from the trades unionists all over the country. From a day of small things they had raised up a band of 70,000 men in the Labourers' Union. He was also happy to inform them, notwithstanding that, at the first outset, two hundred farmers and clergymen met together to crush the movement, they had supporters on every side — members of Parliament, magistrates, landowners, and parsons, but very few of the last. He was inclined to think that the consideration of the question must produce some good for the labourer, for whether in the railway carriage or the drawing-room, he heard the question discussed, and even in the rooms of the Farmers' Protection Association he heard them discuss the best means of helping the farm labourer without giving him anything. He thought the labourers had too long looked for help in the shape of gifts; but they had looked so long that they had become altogether passive, because they had looked in vain ; and he thought it would be a great deal better for them if they could be taught the great lesson of self-help, and the nobler one of helping one another. Talking to an audience of trades unionists, he need not tell them of the advantages of union, but he knew full well that the weakness of the farm labourer was owing to his having stood alone, and even having opposed his brother. He then went on to speak of the true condition of the labourer, saying he was quite convinced that if the men living in towns could actually and truly realise the condition of the agricultural labourer, there would not be many rest easily that night, it was so deplorable. When speaking to some persons of the paltry earnings of the labourers, he was told that they often came in for perquisites ; but he could assure them they were very poor indeed. He described the various perquisites of the farm labourers, as ascertained by careful inquiry, and said he could assure them that, as a rule, the farm labourers had no more given them in addition to their weekly earnings than the North farmers gave to their labourers in this locality. He spoke of the low rate of wages given by the farmers in the southern and western counties, and said that in many instances, when the men had dared to join the Union, they had been turned out by hundreds. Speaking of the opposition the movement experienced at the hands of the clergy, he said they wanted a Christianity which could be practised by the preachers. He believed they had a right to expect to be happy on earth as well as hereafter, and he said, Away with such trash as that doctrine which opposed such a thing. However, the movement had done some good to the clergy, and they would be glad, perhaps surprised, to hear that in Norfolk the Union and its influences had caused the parsons to work. The labourers had been locked out by a farmer, and the clergyman went to the farmer and assisted him in feeding the pigs. He was prone to think if the Union could get the clergy up to pig-feeding they would confer more good upon society than they had done for many a year. Their first word must be "defiance;" they must organise an army to drive the invader from their rights; and when that was done, they would take good care to keep him at a distance. They had not been content with taking the goose from the common, but they had taken the common from the goose; and in Clifton he found that a charity which had been left for the poor for two hundred years had by some means got into a wrong channel. He appealed to them as men knowing the power of organisation, and knowing the necessity of union, for assistance. He need only remind them of the difficulties they had to contend with because of the ignorance of the poor labouring men. He thought they wanted more men to lead the people rather than to drag them back. This was an age of advancement, and it almost depended upon them to decide whether these poor men should be left in the miserable and degraded condition in which they were to be found that day. On account of continued lock-outs, hundreds of men had been removed weekly, and that involved the expenditure of a deal of money. A lock-out having taken place, the men would starve if fresh work were not provided for them, and emigration had to be resorted to. It was a necessity, and, much as he deplored it, he believed it must be done. He concluded by announcing that a committee had been formed to raise subscriptions, and to send them to the central society, and all subscriptions would be gladly received at No. 15, Houndsgate.

The resolution was then put to the meeting and carried without a dissentient, amidst great applause.

Mr. D. W. Heath, who was well received, said he rose to propose a vote of thanks to the Chairman for his kindness in presiding over the meeting. He hoped the vote they had just given would not rest where it was recorded, but that they would put their hands into their pockets, and individually and throughout their various trade societies he trusted they would week by week send large subscriptions to the central fund. By so doing they would set a noble example to the great industrial centres throughout the country and help those who meant to win a great and triumphant victory. Let it be a war of right against wrong, and let nothing stand between them and the union of the agricultural labourers;— let it be a war against that one who was preventing the labourer from earning that which would make his home happy and relieve his wants, and who was preventing him becoming a happier and a better man. He asked for their hearty support, for the support of the cause of labour and right against dictation and wrong.

The resolution, having been seconded, was carried unanimously.

Three cheers for the Union, and three cheers for Mr. Cox, concluded the proceedings.

A UNION DELEGATE'S SPEECH.

Almost all the agricultural labourers of Redburn (Bedfordshire) appeared to be assembled with their sons and some of their wives on Redburn Common, on Thursday, April 24th, when a delegate of the National Union and several gentlemen from Luton spoke on the agricultural labour question. For three hours in a bittter, cold wind they stood on the grass, and all, especially the women, listened intently. Mr. M. Judge presided.

Mr. Allington, a delegate from the executive committee of the National Union of Leamington, rose and said—Mr. Chairman, friends, and fellow working men, we are here to-night to talk about a matter which affects the interests and well-being of the general public as well as the interests of agricultural labourers; and I think if people will look at it as they should, on both sides and top and bottom, they will see it is a question affecting the well-being of this country of ours. We want to discuss this matter in a fair and honourable way, and if there be any farmers in this meeting—and I hope there are—we are not, I hope, the enemies of the farmers. If they can confute my arguments, they shall have a fair and patient hearing, and I will reply. Now, what is the agricultural labourer's position? They may be paid 9s., 10s., or 11s. a week. In Dorsetshire, where I started a union, there were men receiving 9s. a week, some 8s., and a few, a very few, 7s. a week. In other counties you will find men receiving 10s., 11s., 12s., 15s., 16s., up to 18s. and a pound a week, some 22s. in Yorkshire. But, taking the general aspect of the labourer's condition, is it what it ought to be? Does the condition of the labourers on the soil reflect any credit on our country? If not, then I think it is one which honest, industrious, agricultural labourers have a just right to complain of and to mourn. What is to be done to effect a better state of things? Some said, let the labourers go to their employers themselves, not let middle men come in. I have been a labourer ever since I was eight years of age, and I know the condition and difficulties of the working men. All your life long you may toil, but in old age nothing but the workhouse stares a labourer in the face, and at last they will put him into a pauper's coffin and a pauper's grave, while others have reaped the benefit of his industry. When workmen have gone to their employers without intervention of middle men, what has happened? It generally happens that there is one man on a farm who has to speak to the master for himself and the other men. With regret I say it, the farmers generally have not had the honour to call their men together and discuss the matter respectfully, and to say to the men—"I wish to strike a just bargain with you, and I will pay you a fair wage so that you can live—live in comfort." No; but the man that asked for the means to live has been singled out in thousands of instances. I could give names, places, circumstances, where they have tried even to get other men turned out of their employment and out of their cottages. Seeing that such conduct is perpetrated towards working men, what are they to do? With the present price of commodities, what are we to do if we are to maintain ourselves and our families respectably? When men have used every legitimate means to induce farmers to give them a fair wage for their industry, and cannot succeed, what are they to do? The greatest obstacle to the rise of the agricultural labourers is their ignorance—ignorance of their own power when united in one strong compact body. We have been taunted with our ignorance, but whose fault is it that we are so? Our parents never received enough money to enable them to send us to school; and yet the very people who kept them down have said we are an ignorant class of men. That, I think, is cruel in the extreme. There is as honest a heart in the breast of agricultural labourers as in those of the farmers; and give the labourers the means of living in comfort, and they will feel as much interest in the education of their children as the farmers take in that of their sons. There is as honest a heart beating in the breast of the labourer as in the breast of a mechanic; and he has as much regard for his wife and his family as any other man in Old England. Is there a man here who will say that twelve shillings a-week is enough to support a man and his wife and family as they ought to be, taking into consideration the price of everything we have to buy? These are the questions gentlemen should ask themselves; and if they think a man cannot live with it, then they ought not to meet to form associations to crush a union which means directly to benefit the lower order of society, and to injure no one above it.

Ought they to do it? The day is dawning, my friends, when the working men of England will not submit to the tyranny which has been brought to bear upon them in the past. The unions of mechanics have helped a great deal in educating the minds of workmen to the power of union. Trades unions were the starting-point towards the emancipation of the British labourer. I am thankful for penny papers which circulate in the country. They are powerful engines for the diffusion of useful information; and in these papers, selling by thousands, the working classes are taught that they have a right to live. A farmer in Warwickshire said to me, when I was talking about the desirability of labourers' children being educated, and saying it was a disgusting thing, when a man had four or five children, that he must send the little boys, almost as soon as they began to walk, into the fields to scare crows, or tend pigs, or drive horses, depriving them of that education which God intended they should have, for if God gave us intellects, He intended they should be cultivated and developed,—this farmer said—"What do the labourers' children want education for? Enough if they can write their names and count twenty." We do not want to have recourse to harsh speeches to carry on a noble cause like ours; but I say it with all due respect to farmers—there are honourable exceptions—but the great proportion of them have entertained the same sentiments respecting the education of our children as this farmer said to me—" Keep them in ignorance if you want to keep them down." It is said that if the men have more money they will squander it away at the public-house. How do they know that? They only suppose such a thing. I am bound to say it, from an honest conviction, that there are as moderate men, as temperate, moral, and kind-hearted men, wearing fustian jackets and billycock hats, as there are amongst the farmers. We say we have a right to live, seeing that we are the producers of the wealth of this country.

Great stress is laid upon capital, and it has its rights. But then intellect has its claims too, and labour has its rights. I never knew a man who managed his land properly who ever became bankrupt; if he did, he was not an economical man, but a spendthrift. Seeing, then, that as working men there are no other means whereby we can obtain rights only by uniting together, I ask what are the objects for which we unite together? It was said at the farmers' association at Dunstable that we were banded together, not to make men better, but to make them worse. I deny a statement like that. I had the paper put into my hands at Bedford, and I am prepared to state that they uttered some of the falsest statements that ever escaped the lips of mortal men. A cause which needs to be advocated by untruths is not honest, just, or fair. It was said that were the men thrown on the Union funds there would not be enough to find them a breakfast. The man who said that, mark you, was more liberal then than farmers are on pay-nights. He said we had only funds in hand to the amount of 6d. a-head, and that would not suffice for a breakfast. Why, you men don't get a penny a-head per meal for yourselves and families to subsist on. I will confute his statement. The Union has its thousands, and we are increasing every week financially, numerically, and in power and influence throughout the country. I will defy the farmers to crush the Agricultural Labourers' Union, because our cause is founded in righteousness, and we aim to elevate the down-trodden sons of toil. What has the Union done? It has done more for the agricultural labourers of England than all the parsons that have preached in the Church of England yet. Yes, sir; the teaching of the ministers of the Gospel—and I don't confine my remarks to one section of them only—is contentment in poverty. Do you question that? Look you here; when a man who works day after day cannot live on his industry, he has a right to make known to his spiritual teacher his circumstances in life; and as the overlooker of that man, I might say both as to body and soul, that teacher has no right to say a man should be contented who has been robbed of his rights. The ignorance of the working classes has been such, they have thought they ought to work and support the minister, whilst they themselves have had to live in a wretched and miserable condition, and they have been told that these things are in the order of Divine Providence. That is one of the most blasphemous doctrines ever taught by man. God never created man to starve him to death; I challenge the parsons to prove it. When your spiritual teachers teach you a spurious doctrine like that, I have a right to tell you that you have a just claim to live by labour. The Agricultural Labourers' Union is doing a vast amount of good to others besides labourers. Let but the labourer take more money, and the shopkeeper, the butcher, the baker, the grocer will know of it. Honest men have been taught to be contented in that state of life in which it has pleased God to call them. The day is coming when teaching like that won't do. Let us have honest teaching; teach the farmer his duties, and the labourer his, and be just, honest, and fair in the sight of all men. It was stated in the London "Times" that last year an increase of one million was paid to agricultural labourers. Could that have taken place without the Union? And has it not done the men and their wives and little ones good? We hear a great deal about capital. But is not labour capital? Does gold till the fields, gather in the crops, and thrash the corn? You know better. It is the strong arm of the English labourer which does it. I have heard since I came here that they are trying to buy you over with little bits of land. Don't you be bought in that way. Demand to be paid for your labour in the current coin of the realm, and have your cottages direct of the landlord. When we first started a union in Warwickshire, the farmers discharged two hundred men. We sat up night after night writing appeals, and in six weeks we received £800. Don't you fear the farmers. We have more money in our exchequer to-day than we ever had before, and it is continually increasing. We don't want to crush the farmer, and we are not going to let the farmer crush you. Many a farmer has been compelled to give up his farm as the consequence of joining farmers' associations and their opposition to the Labourers' Union. If farmers here won't pay you a just wage, migrate, and emigrate to where you can share in the prosperity of the country.

Addresses were also delivered by Mr. Smith, of Luton, Mr. Paul, of Sandridge, and others; and a branch of the Union was formed, a treasurer, secretary, and committee appointed, and a number of names taken down.

It was stated that a shepherd six years on the farm of Mr. Olney, of Byelands, had got notice to leave. The man had been taking 13s. a-week: he has a family of seven children. He lives in a cottage on the farm, and it was expected he must turn out with his family by Monday next.

GREAT MEETING AT NEWBURY.

A great demonstration was made at Newbury on Friday evening, August 18th. Members of the village branches came into the town in large numbers, and partook of tea in the Fair Close. They subsequently perambulated the borough, and at seven o'clock a meeting was held in the Close, and much enthusiasm prevailed, quite an ovation being accorded to Messrs. Arch and Taylor. Mr. Councillor Lucas was voted to the chair, and after a short speech, approving of the Union, introduced

Mr. H. Taylor, General Secretary, who, in the course of an able speech, said there was probably no movement being carried on in the present day which had caused so much stir, and yet had been so little understood, as that being conducted by the National Agricultural Labourers' Union. He felt great pleasure in seeing in the chair an independent gentleman, who declared himself an advocate of a fair field and no favour. The representatives of the Union had not come to Newbury to try to thrust their dogmas down people's throats. What they wanted to do was to teach the law of the people, and instruct them as to what the Union proposed to do. If people who read their rules, watched their progress, and strove to understand their aims and intentions, afterwards came forward and honestly declared they objected to the movement, they could shake such by the hand as honest opponents. But what the friends of the Union complained of was, that before men had even thought about this movement, or read the rules, they acted like the bull at the haystack—ran at them, and tried to pitch them about. He would say such conduct was indicative of the spirit of the fight. He mourned to see a person occupying the highest municipal post in the town so narrow-minded, and so ignorant of what was going on around him, as to prejudge a movement like this, and use such arguments, if arguments, indeed, they could be called. He regretted to see the Mayor of Newbury had said, according to the report of a local paper, that he objected to this movement because it set class against class. He would ask the Mayor of Newbury or any one else, whether it was not because class was already set against class that they had met there that day? It was because they had found that the farmers and their men, whose interests should be mutual, were fighting against each other, that the Union were anxious, if possible, to reconcile them to each other; but so long as the farmers were determined to pay bad wages, and only about half the value of labour, they must expect there will be an antagonism between the classes. He congratulated his audience that the starting of this movement had resulted in bringing the eyes of the country upon the agricultural labourers. Whether he went into a railway carriage, or the drawing room, whether he went to the meeting of the Farmers' Defence Association, or elsewhere, he found the clergy, the landowners, and the farmers considering the best means of elevating the farm labourers, but without giving them anything. He went on to point out that the labourers had in their own hands the power of helping themselves, and begged them, that instead of working against each other, as had been the case too often in the past, they should stand by each other. He gave the men excellent advice as to improving their social habits, and after regarding the Union movement from various points of view, spoke of its vast and increasing extent, and called upon the labourers to support its good work by enrolling themselves in its ranks.

Mr. Joseph Arch was announced as the next speaker, and was received with great enthusiasm. He addressed the meeting as "fellow countrymen," and proceeded to express the pleasure he had in being present at Newbury to discuss a question upon which hinged the weal or woe of England. A great many people who happened to be the fortunate possessors of wealth had no sympathy with the movement; but this was a great mistake which had been in existence too long, and must be rectified for the dignity, prosperity, and safety of the country. As a labourer himself, he felt pleased in being able to advocate the cause of his fellow-men, and better their condition. If he was doing a work that did not require being done, or doing it in an un-English, unlawful manner, he was prepared to receive instruction from any minister or gentleman in that audience, if he would kindly come upon that platform and teach him a better way. If he felt there was a wrong which ought to be rectified, and the exposure of that wrong would bring down on him the frown of the whole country, then he would raise his voice against that wrong, if princes and peers condemned. Whether he preached to the poor agricultural labourer, or addressed the prince or the peer, he preached alike to one and all the same grand old Bible doctrine :—

> " Let Cæsar's due be ever paid
> To Cæsar and his throne ;
> But consciences and souls were made
> To bow to God alone."

From earliest infancy he was taught by a sainted mother that his creation in the world was to be a man, and never a slave. He might tell them, however, that he knew something of the chains of a slave. When as a father and a husband he had to work for 1s. 6d. a day, and was expected to maintain a wife and family creditably, he found he could not, and felt that he who would be free must himself strike the blow. More than twenty-five years ago he struck against the deadly monster oppression, and from that day to the present he had never bowed a single sinew in his neck to that tyrant. The same steps that he took each of them must take, and they must make this a text from which their everyday life must be preached, that if they helped themselves then God would help them. He was not going into the grievances and sorrows of the farm labourers of England, because, thank God, the land serfs of England were getting their eyes opened, and all the landlordism of the country would never force them back to where they were. The public had their eye upon the Parliament now sitting; they found that the Education Bill was all but, if not quite, abortive; that the Arbitration Bill was to be treated "as it was in the beginning is now, and ever shall be ;" and that Government would take the same course, and fight when it pleased. These were grave topics, affecting labourers, farmers, shopkeepers, and all of them, and yet they were treated as of no importance. Nor was the conduct of their present Mayor calculated to inspire a good feeling amongst the working men of the town and district of Newbury. Not that he would have liked to have spoken in the Corn Exchange, because he would rather speak where he was ; but he would have liked that gentleman to have shown a good feeling towards this movement. He would go to the feet of no mayor, public officer, farmer, or landlord, while Almighty God stretched above their heads the broad canopy of the bright blue sky, and under their feet the green sward ; and those of them who had been accustomed to what some people called ranting preaching, were prepared to take their stand where thin-skinned gentlemen would feel a delicacy in doing. He thought no public officer should pander to one class of the community. Their public buildings should be free alike to all, whether the politician, the clergyman, or the labourer. To shut one class out from the privilege which was granted to another class, was the way to set class against class. The Mayor has denied us the Corn Exchange, but our heavenly Father has sent us a beautiful nice fine evening, and let us have this spacious building. He wanted them to look at some of the grave points arising out of this movement. He might have some farmers present. He once addressed six hundred farmers at Dorchester fair, and a most attentive audience they were. He had not been a farm labourer all the days of his life without knowing some of the difficulties of the farmer, and he would defy any master that ever employed him to say that he (the speaker) did not take an interest in his work. That interest which he had always felt for the farmer was more intensified than ever. He found that one-fourth of the farmers had no security whatever for the investment of any capital. A farmers' paper said there were four hundred farmers nearly bankrupt, and four hundred more who exist only by the good will of their landlord. If this was a true picture, then the farming interest was the most depressed and insecure of any great industrial branch in the kingdom. Some of the farmers thought, when Messrs. Howard and Read brought in their Tenant Right Bill, that they would soon be let into a good thing. But they might depend on it that they would never get it until their labourers had the franchise; and even then not until the farmers had pluck enough to go into Parliament and represent their own interests. Scotland, at the next election, meant to send ten tenant farmers to the House of Commons: how many did they mean to send from Berkshire? Farmers had the ballot; dared they use it? Farmers had political power; what had they done with it? Why, crawled snail-like to the feet of the squire, and voted as they were told. They sent gentlemen to Parliament who fettered and bound the farmer with chains, and then screwed it out of the labourers to make up for any loss. He told them candidly that time was gone by; they must drop that dodge, or else it would be too hot for them. He told farmers that their wisest plan was to treat their labourers as men ought to be treated, to pay them as men ought to be paid; and if they had made a mistake by taking their land too high, and let landlords shoot game which their tenants had fattened but did not get a taste of themselves, then he warned them that if they made up for their mistakes by starving their labourers, they must drop that dodge, or, as labourers, they would be off to another land with their labour. He might now be allowed to say a word or two to the landlords, many of whom, like Lord Leicester, had said, "Build the labourers good cottages, and let them at a moderate rent." But then Lord Denbigh had said, "You must stop in them, and work for the farmers in your parish, or you will leave my cottage." Did they think any reasonable farmer would take Lord Denbigh's farm, if he introduced a restriction in the agreement that the beef and mutton and hay and corn on that farm were all to be taken to Leicester market, though perhaps the farmer might get better terms at Nottingham? And did they think labourers were such fools as to take their cottages on those terms? If the landlords adopted that dodge, there was no downright intelligent, plucky English labourer who would be trodden down in that way; for if that was the slavery of his own country, he would leave it and go to America, where he could find freedom and fair play. Last year 300,000 emigrated from our shores, representing forty millions worth of bones, sinew, and muscle, and the stream of emigration was going on; and when the best were gone, the thrifty and most intelligent, those left behind would be the weak, infirm, and those within a few hours of pauperism: and who would have to pay the rents and rates but the farmers? A farmer once said, while he could find money he could always find men. Towards the latter end of August he was going to the States of America. If he found that there were farmers who bought and sold, and had both sides of the bargain, he would say, "Stop where you are, chaps, the crows are as black here as they are in England." If he found that country the home of the working man, where the labourer was free to make his own terms, if his boy could sit down on the same form with the boy whose father had got wealth, read out of the same book, and write on the same slate, where the poor man had political power the same as the classes above him; then, if the farmers would not treat their labourers like men, if they would follow him he would lead working men across the broad Atlantic to the fruitful fields of America, with its ninety millions of acres yet untilled—he would stand upon the shores of America until he had drained the resources of England, and had made the farmers bite the dust. He was not shelving the question nor muzzling it. He had appealed to the farmers from the various platforms of England, like as an Englishman would. He had advocated their rights as much as the rights of their labourers, and what had he met with from their class but censure and attempts to ruin his reputation? They had tried to put their iron heel on the movement, which had from the first been law-abiding and as lawful as any other institution in the country. They had locked the men out by scores and hundreds in different parts, until he had been compelled to raise £2000 from the public to keep those men and their families from starving. They were guilty before God and man, and if they began to play this dodge again he would make them before this day twelvemonth know the worth of a man. He would treat them with the utmost courtesy and friendship. He had fought their battle before the Game Committee of the House of Commons, and they knew it. He had been in company at private dinners with noblemen, and had depicted farmers' wrongs as vividly as ever he had depicted the wrongs of the labourers. He had shown the great landlords of the country that they were doing the farmers an injustice as a class, and yet, in the face of his humble advocacy and these honest statements, they had tried to put their heel on the neck of Arch and the other men whom he had the honour to belong to; but if they came that dodge again, he would make them know the worth of a man —they should have the fields, but not the men to till them. They put him on his mettle, but he was not to be frightened by a trifle. Some of the farmers in his own neighbourhood said the Union would be only a nine days' wonder, but one of them who knew him from a lad, said, "If Arch has got anything to do with it, before you can make him loose his hold, you will have to cut his head off." Though an agricultural labourer and a man of humble birth, he had English blood in his veins, and an English heart in his body, and he was prompted in what he did by human feeling. When he saw 600,000 tillers of the soil in slavery, and mocked by being told they were free, he would traverse America, if he lost his life, in trying to raise his fellow-labourers—he would die gladly if he could only bring them up to freedom. But now a few words about the clergy. The Bishop of Gloucester said to a friend of his, "I had thought that Arch was a mild sort of man, but he does speak some very strong things." Well, if he did not tell the truth, he was willing to be corrected. The question had cropped up in the House of Commons, on which Mr. Miall had the bad luck to be defeated—he referred to the Disendowment and Disestablishment Bill. He was not about to give them his own private opinion, but viewing it from a political standpoint, and, what was better still, a Bible standpoint, he had no hesitation in saying that that Church which had not got enough of the love of God beating in the hearts of its members to support it, and keep it alive without being propped up by the State, then let it go. He would not be a member of such a Church, nor a parson in such a Church, for a good lot. He stood before them that night a Methodist local preacher of twenty-five years' standing. During that time he had walked more than 7000 miles on his own conveyance to preach to his fellow-men. He had never had a sixpence from the State for it, nor did he want it. He knew something of the trials to which a minister of the gospel was exposed. No man would more liberally — and there were gentlemen in this meeting who would confirm this statement—of the little God had given him contribute to the comfort and happiness of the earnest minister. No man more than himself would wish to see a minister of the gospel free from life's cares, and the difficulties which would press him down, and make him weep and sigh in his retired moments. He himself had gone eighteen or twenty miles out to an appointment of a Sunday with sorrow and grief pressing down his heart, and what was the cause for that sorrow? Because he could not pay his way in the world. Why was it that he had not enough to pay his way? It was not because he spent his money in drink, because for seventeen years he had been a total abstainer. He had drunk the bitters, and he would wish no Christian minister to drink the bitter dregs he had. When, however, the minister of the gospel left the performance of his holy functions, and turned aside to serve tables, and became a money-monger, or one of Her Majesty's Commissioners of the Peace to send poor women to prison, then let them have him out of it. Unless Church of England ministers protest against clergymen holding commissions of the peace, they would have repetitions of that which so many of them delighted to indulge in, the wreaking of vengeance on the poor of their parish—they had wreaked it on his head many a time because he was a Dissenter. When the State found out that men with no better religion than that which wreaked vengeance upon the head of the poor and helpless occupied its pulpits, then the State must have that church upset, or the State would be called to account. When a clergyman, because the squire sends him a brace of pheasants at Christmas, invites him to his splendid table, and sends him bottles of wine, panders to the man of wealth and shuns the house of the poor man, it was a disgrace to Christianity; he was a traitor to his charge, and he (the speaker) would not stand condemned before God for a world. Canon Gregory, beneath the dome of St. Paul's, said that this country was approaching a serious climax. One class was getting immensely rich, and building up great colossal fortunes; others were extremely poor, and all but starving. This it was that brought Rome to the dust and made the Emperor of France turn up his sword to the

Prussian powers. Something must speedily be done. They must send men of business to Parliament—men who will not snore like Berkshire hogs—if the sceptre of the country was to be wielded with judgment. There were far more important questions that demanded attention than the Shah of Persia. There were Englishmen who had a right to live in the country, but oppression said, "You shan't, we'll drive you from our shores," and when their legislators began to break the grand moral law, they must take the consequences; if they began a row they might fight themselves, and if they wanted property defended they might defend it. Place the labourer in a free market and in happy situations, invest him with his rights, give him a moderate stake in the soil, two or three acres to till for themselves; let the landlords, labourers, and farmers shake hands, pull in one boat, and have an eye to each other's interests, then let a foreign foe put a foot on this country, and there will be no want of defenders, for no men make better soldiers than the English labourers. If they were going to send the best men away from the country—for emigration meant grubbing up and taking the young stock with them—where were they to get their future supply of labour from? This might be a matter of indifference to some who were pocket-proud—who had more money than sense. Some people who had a few pounds in their pocket were as cocksy and thought themselves better than others; but it was all bosh and nonsense; they were only flesh and blood the same as others, and if Providence had made their path smoother, then it was their duty to help their poor and struggling brother. They might fancy that their mountain of wealth and property stood strong; but if they lost the tillers of the soil their country would decline and decay. They might think that it was by Divine appointment that one man had money and another was poor, but God would teach them a different lesson. He would teach them that human life and souls were valued by Him, and if they began to value property more, He would take it from them; He would send a scourge upon the country. God would not be mocked by a defective Christianity which taught men that they might rob their fellow man, starve and impoverish him, and that it was all in the right of divine appointment. As a nation they had mocked God and set Him at defiance. He had stayed the sword, and up to 1873, God had conquered their foes. But let their labourers go from her, their army desert and lose confidence in royalty, and people begin to ask for a republic—and who said they would not—what would happen then? Much as he loved his country, if he saw within her men who were slaves, men who were allowed to pine nearly to death, men who were denied the common necessaries of life, and yet worked and toiled hard for the millions above them—he would raise his voice and use his influence on both sides the Atlantic till England should consider her ways and treat her industrial population as she ought, and no man would more delight to see stamped on her insignia that which she should ever carry, "the pride of the ocean, the first gem of the sea."

The "Newbury Weekly News" says:—"Mr. Arch may be described as being of medium height, and what is known in popular terms as a 'thick set man.' He wore a billy-cock hat, a short, round jacket with pockets at the side, and presented the appearance of an ordinary labourer in Sunday attire. He spoke in the vernacular common to a Warwickshire peasant, but with a command that not a word was lost, nor an interruption occurred from the vast company which was closely packed, and hung with bated breath on the lips of the speaker for more than an hour, only relieving themselves by the applause following each sentence, which was delivered with more than usual animation. Whether for good or for evil, he has probably a greater influence over an out-door assembly of labourers than any other man in England, and an impartial observer, judging from the impressions of Friday night, would be of opinion that as long as the movement has such a man for its champion, there is little doubt of its vitality or even aggressiveness."

These necessarily imperfect sketches of the *modus operandi* of the Union, will perhaps be the better understood if supplemented by a few incidents of rural life. One of the great bugbears of the agricultural labourer is the parish workhouse. It is probably one of the most depressing circumstances of his not particularly sunny existence — the ever-present possibility of ending his days therein. The following extract from the "Day of Rest" furnishes a graphic picture of what it is in the "house," as it is universally called, which specially causes it to be so much dreaded.

A FAMILIAR WORKHOUSE SCENE.

We stood inside the great black doors, which swung to behind us, shutting us in as though they would never open again, save, may-be, when we were borne out through them in a pauper coffin. Transome leaned more heavily on my arm. A man in the workhouse suit was sitting in a little room just within the doors, and as we stood staring about us he called out sharply :—

" Na, then ! whatten yo standin' there for? Canna' yo come on and tell me whatten yo want here ?"

" Me and my husband have brought an order to go into the House," I said.

" Inside birds, eh ?" said he, laughing a little; "caught and caged! Go on, then, t' th' measter's office. First dur t' th' reet across th' yard."

I guided poor Transome across a large, square yard, with nought to be seen save high walls on every side, with windows in them that had no curtains, like eyes without eyelids looking down on us. But there was not a face to be seen at any of them, and a mournful stillness filled the place. It was Transome that knocked at the master's door, a quiet, feeble knock that could never have been heard if there had been much noise. We were called to go in, but we did not stay there many minutes, and the master sent a man with us to show us our separate wards.

Once more we had to cross the great yard, Transome clinging to my arm, till we came to a door in the wall, where we must say good-bye to one another. We never had said good-bye all those long years, those forty years since he had taken me from my father's home in another county. How could I let him go out of my sight? It was not like him setting off for his day's work, sure of coming in again in the evening. How could him and me spend our time apart?

"Could na yo' leave us for two or three minutes?" said Transome to the man, feebly. "Hoo's been th' best wife as ever a man had these forty years; and aw dunno how to bid her good-bye. Give us a minute longer to be together."

"That aw will," answered the man. "But it canna be more nor a two or three minutes. Bless yo! yo'll see one another at prayers morn and neet if yo chosen to go; and yo'll ha' half-an-hour o' Sunday, besides half-a-day out once a month. It's noan so bad is th' house, so as yo' getten reet side o' th' measter."

He went off for a little while, leaving Transome and me against the door into the women's wards; with all those dark, staring windows looking down upon us. I laid my head against the door-post, and broke out into heavy, heavy sobs.

"Na, Ally," cried Transome, "na, my lass! Hush thee, hush thee! God A'mighty's here as well as out yonder i' th' world. He knows where we are; and sure He loves us both, same as He's loved us all along. We mun put our trust in Him, and go through it: thee and me mun part. Eh! but aw wonder if God A'mighty looks down on ony heart sorer nor ours at this moment o' time?"

"Only promise," I said, through my sobbing, "promise me faithfully you'll be careful of yourself, and keep up, so as we can get out again in spring, when the warm weather is come. Oh, Transome, if I could only keep nigh you, and take care of you, I shouldn't mind."

"There's One as'll take care on us both," he answered, his voice trembling; "One as says, 'I'll never leave thee, nor forsake thee.' O'ny think o' that, my lass. He's here i' th' workhouse itsen; and nought'll part Him away from thee nor me. Good-bye, Ally. Aw hear th' man coming back to us."

He stretched out both his hands to me, and I put mine into them, and we kissed each other solemnly, as if we were both about to die and enter into another world. I saw his face quiver all over, and then there came across it a patient and quiet look, which never left it again, never! I knocked at the door before me, and passed in, just catching a last sight of him turning away with nobody to lean upon. Then the door was thrust to between us, and I could see him no more.

I did not heed much what was said to me, and I did not look about my new dwelling-place; only I followed a woman who passed through many rooms, where the windows were high up in the walls, so that nobody could reach the sills, and where there were groups of women all dressed alike, chattering most of them; and there was a strange close smell. Oh! how different from the sweet air in our old home. At last, when I came to myself, as it were, I found I was sitting on a chair at the head of a little narrow bed, in a long room, with two long rows of beds down the sides of it, and a narrow path up the middle. All the beds were alike, and the bare, white-washed walls closed us in, with nothing to be seen through the high windows, save a little bit of grey November sky. There were old women all round me; some of them many years older than me, even a few of them bed-ridden; but they seemed too dull to take any notice of me, as if everything that was like life had died out of them, save the bare life itself.

Well, there's no need to tell you much about the workhouse. Most poor folks know more of it than they care to know, either through their own troubles or the troubles of their friends. I don't say a word against it; only I could not be with Transome. There, think what it was to have been his wife forty years, with scarcely a brangle between us, and never a sulking quarrel, and all at once to be shut up in different parts of the same building, with only a few walls and yards to part us, yet not be able to see him, or even send a loving message to him. I wet my pillow with my tears that night; ay, more than when my Willie died, as I wondered and wondered how he was faring, and if he was warmly wrapped up, and how his pains were. But I could do nothing for him, no more than if I was lying in my shroud and coffin. At last my loneliness and my trouble drove me to remember Him that is everywhere, and was with Transome as He was with me. "Lord," I said in my heart, for it was not altogether a prayer such as I had generally said to Him, "Lord, if they'd only make his bed comfortable, and wrap him up well in the blankets. Do put it into their hearts, Lord, for he's tried to serve Thee faithful all his life long."

After that I felt a little easier in my mind; I fell asleep, and dreaming of the days when Willie was alive, only sometimes the child was Willie, and sometimes Pippin. I suppose it was because I had close to my pillow the little box that held the curl of Willie's hair, and Pippin's piece of money. It was the only thing I had brought in with me, except a few bits of linen Transome had woven for me years and years ago, which I had bleached as white as snow in the frost on the brow of the hill.

My next incident illustrates a hardship which, owing to the discretionary powers vested in rural guardians, enables them to visit with something very much like a fine, the offence of Unionism. The case is no uncommon one; but as this one was taken up by the Leamington committee, I give the following report of it, as found in the pages of the "Chronicle":—

THE POOR-LAW GUARDIANS AND THE LABOURERS.

THE MAINTENANCE OF AGED PARENTS BY THEIR CHILDREN.

Southam Petty Sessions, Warwickshire, June 16th, before Henry Thomas Chamberlayne and A. H. Thursby, Esqrs. Walter Mawby *v.* Joseph Overton. The complainant was the clerk to the guardians of the poor of the Southam Union, in the counties of Warwick and Northampton, and Joseph Overton is a labourer at Harbury, who was defended by Mr. Edwards-Wood, of Leamington, solicitor. The complainant's case, proved only by his own testimony, was that Thomas Overton and Fanny Overton were poor old and impotent persons, not able to work, and had become chargeable to the Union on the 2nd instant, and that the said Joseph Overton was their son, and of sufficient ability, at his own charge, to relieve and maintain his said parents. —Mr. Edwards-Wood then addressed the Bench at much length, and brought out the facts in bold relief that the two pitiable paupers, man and wife, had been tenants of the chairman of that honourable bench for forty years, in the parish of Ufton; that no couple had ever passed through life more creditably than they; Joseph Overton having worked forty years for the trustees of the turnpike, brought up most respectably a family of four sons and two daughters, and the misfortune of receiving parochial relief never befel his lot until the 2nd of this month. And as the advocate of the son, he had the pride of saying every one of the said children had followed in the honest footsteps of their father, each of them being married, and supporting their respective families by the fruits of honest, sober industry. Fortunately for his client every circumstance in this case was within the personal knowledge of the chairman of this Bench, and who was also the chairman of the Union, so that he hoped justice would ultimately be done between all parties. As regarded the statute under which the claim was made (43 Eliz., cap 2), he contended that it was never intended to compel paupers to support paupers, although they might be blood relations. In order for the son to become liable to do so, the condition laid down in the Act was this, "he being of sufficient ability." He had carefully made out this labourer's balance-sheet, who was present to swear to the truth of every item it contained, which showed that with the present dearness of every article he had to buy, he had to meet a deficiency of £7 11s. 6d., instead of having any surplus to meet the demands of the Southam Union—which showed the injustice and inhumanity of the guardians' procedure, which attempted to get rid of a legitimate liability on their income, by removing it to the shoulders of a son who had not sufficient to meet the daily wants of himself and his family. The ability to pay had been the pivot on which all the highest judges of the land had rested their judgments in every case which had been submitted to them. He quoted as above all decisions the express words in the Act. Then came to our further guidance the case of Rex v. Reeve, Michaelmas Term, 7th Charles 1st, before Mr. Justice Jones and Lord Chief Justice Croke, whose significant words were, " It is very reasonable that he being of sufficient ability, should contribute to the support of his grandchild." The like principle was acted upon in the Jew case of St. Andrews Vindorshaft v. Jacob Mendez de Breda. " He, being very rich," had been ordered to pay 20s. per month for his daughter's support. That was the law in Michaelmas Term 13th William 3rd, when that appeal was heard. Mr. Justice Eyre made the like remark in the case of Shormanbury v. Bolney, 5th William 3rd. In the case of of Rex v. Munden, 1st George 1st, it was laid down that the assistance must be according to the words and intent of the Act. And in the case of Woodford v. Lilburn, 20th George 2nd, it is stated that the case of Rex v. Munday settled it to be for blood relations, and that even those were not liable "unless they be such as are of sufficient ability." Therefore he considered the present a most iniquitous and unrighteous attempt to reduce his client to be ultimately a candidate to have workhouse relief for himself. The balance sheet of the defendant was then discussed, and the clerk of the Union was unable to point out any objectionable item in the expenditure.

The balance-sheet was as follows :—

RECEIPTS.

	£	s.	d.
Wages at 15s. per week	39	0	0
Extra 5s. per week during harvest month	1	0	0
Profits from allotment :—Sack of wheat 18s., potatoes 10s., beans 15s., straw 15s.	2	18	0
Extra 3s. per week for hiring month	0	12	0
Value of vegetables from allotment and garden the year round	2	10	0
	46	0	0

		£	s.	d.
Less deduction for three weeks and five days of lost time, being one day per fortnight		2	17	6
Total income		43	2	6

PAYMENTS PER WEEK.

	£	s.	d.
Rent	0	2	0
Coals	0	1	6
Bread, 5 loaves at 7½d. per loaf	0	3	1½
Rent for potato ground	0	0	7
Grocery	0	3	0
One gallon of flour for puddings	0	0	7
Pigs, cost post price £2 and £1, per week	0	1	1½
Feeding the two pigs (barley meal 17s., stuffs 11s. 6d., beans, &c.), per week	0	2	6
Club money per week	0	0	6
Wife and one child's ditto per week	0	0	1½
Man's clothing	0	1	6
Wife's and child's clothing, bedding, &c.	0	1	0
Butter, 1lb. per week	0	1	4
Cheese, ½lb. per week	0	0	5
Beer for Sunday's supper	0	0	2½
	0	19	6
Total of the 52 weeks' expenditure, at 19s. 6d. per week	50	14	0
Total income	43	2	6
Deficiency	7	11	6

The Chairman made some very feeling remarks on behalf of this aged couple. He endorsed all Mr. Edwards-Wood had advanced in their favour, and also bore testimony to the good conduct of all their children, by whom they had been assisted until they became married. It had been the practice to levy this kind of contribution from the children of persons chargeable to the union. The Bench therefore ordered a shilling per week in this case, and hoped the opinion of a higher Court would be taken, so as to have the law on such cases indisputably settled.

One of the consequences of the growth of the labourers' movement is the attention bestowed on it by the London press. I have already referred to the really valuable service rendered it in the "Daily News," in its earliest stages, by the graphic pen of a "special correspondent." More recently the "Daily Telegraph" has taken to "looking in" upon the labourer, or, to use a more correct press phrase—"interviewing" him.

In running through the pages of the "Chronicle," I find the following:—

THE COCKNEY AMONG THE RUSTICS.

The "Daily Telegraph" has published some letters from Bucks and elsewhere, "by our own commissioner," which is another name for "a reporter" who runs about in search of impressions, which he throws into black and white as quickly as they are made, and sometimes, apparently, before, and which are of the same intrinsic value as thistledown, which is blown about hither and thither, bearing everywhere thistle crops of superficial sentiment and opinion, instead of the solid corn and wine of earnest thought and true heart feeling. We need scarcely say that our "commissioner" looks down from a pretty considerable elevation upon "John Whopstraw," as he pleases to call an agricultural labourer, in weariness of repeating the more hackneyed name of "Hodge." It is really too bad to judge of the intelligence of a class by what is carelessly talked by the lighter wits who frequent the village ale-house. People of whom the following can be written, however, can scarcely be in so very bad a case intellectually, however deficient they may be accounted when judged by superficial standards, and upon the briefest possible acquaintance.

"On Saturday I was at a place called Little Kemble, which is between Princes Risborough and Aylesbury, and there at the village inn were eight or ten haymakers and teamsters waiting until the train brought the London weekly newspapers. By-the-bye, I may here mention as a curious fact, revealed to me during my brief sojourn amongst the rustic natives of Warwickshire and Buckinghamshire, that they have but a limited amount of confidence in daily newspapers. Their objection to intelligence fresh distilled and piping hot is similar to that they have to eating new bread ; and though news a week old may be somewhat stale, it is more solid and satisfying and less likely to ferment. I suspect, however, another reason for the preference—the daily papers are at present 'sour grapes' for Hodge. At all events, the skilled artisans of the hay and turnip fields of Little Kemble were waiting for their 'weekly,' in order that they might get some further confirmation of the amazing rumour. It would have done Mr. Gladstone's heart good could he have seen the eager interest with which the newspaper was spread out to its full length and breadth on the taproom table, while the nine Little Kembleites clubbed their thick heads over it in hopes of lighting on 'soom'at fraash' on the agricultural franchise question. I believe it would not have in the least surprised them had they discovered an entire page of 'parliamentary' devoted to the all-important subject as well as two or three leading articles. They were doomed to disappointment, though they searched with the utmost diligence, and pursued their investigation even to the column devoted to answers to correspondents, in the desperate hope that, though never so tiny a bit, 'soom'at fraash' might turn up there. There was a tremendously long account of a breach of promise case, but to-day, contrary to their wont, the tap-room customers of the Pied Pig had no appetite for that kind of literary diet, and even turned away with an exclamation of discontent from a whole pageful of 'Claimant.' 'Who warnts to read that there kind of stoof? Why don't they gie us summ'at as we keers for?'"

Probably a similar comment would be made by our own readers—"Who warnts to read that ere kind of stoof?"—if we were to reproduce in these columns very much more of the "Daily Telegraph" commissioner's fine writing about agricultural labourers, who, if really possessed of no more intelligence than could be gauged by the Cockney mind, might almost as well be growing trees. Just fancy our all-sufficient Cockney commissioner talking to labourers like this about the difference between Liberalism and Toryism :—

"'Well, for my part,' said I, 'I don't find it easy to see the difference between a Liberal and a Conservative. What is the difference?' And as I spoke I looked round me to discern the village politician who would render me the required information, but to my astonishment there ensued a sheepish silence, most of the company finding it convenient to apply themselves to their beer-jugs. Those who were not so engaged looked hard at the ploughman, who made a desperate endeavour to get over the difficulty by affecting supreme contempt for my ignorance. 'Oi should have thought as how you was old enough to have knowed that without axin'!' said he. 'But,' I persisted, 'what *is* the great difference between the two principles?' 'Why, what's the use of yow a taalkin'?' retorted the ploughman, turning on me as fiercely as though I had been obstinately defending my view of the matter for an hour or more ; 'there's all the difference, I tell yee.' 'What's the meanin' of the Conservatives callin themselves the Opposition, if there beant no difference? What do one soide have blue caards at 'lection toime and t'other side hev yeller, if there's no difference? What makes 'em keep strict to their own public-houses and take to wranglin' and foightin' at the hustin's, if there's no difference? I'm surproised at a man of your years not knowin' better.' And to make my confusion complete the Little Kembleites assembled joined in a chorus of 'Yeer, yeer.'"

This is surely petty, silly work for the special representative of a great daily newspaper to come into the country for. If he had cared to enlighten his listeners and not bewilder them, he might have measured their powers of understanding with far more truthfulness, instead of persistently putting an abstract proposition—What is the great difference between the two principles?—by shaping the question in a livelier mould, as thus: "What is the difference between one person who works energetically to the end of putting things right that are wrong, and of making his fellow man free, healthy, and wise, and another who takes things as he finds them, and leaves them no better;—between 'one who seeks his own in all men's good,' and another who, for his own selfish advantage, would hold his fellows in ignorance and slavery;—between one who has faith in a divine law of right, and in the duty of shaping his own course and that of the State thereby, and another who, abandoning hope and faith, would retread the footprints of the past;—between one who believes in 'Church, crown, and constitution,' and another who has faith in these only in so far as they assist and encourage the hopeful activity of mind and spirit, and tend to the larger development of powers in all, for all;—between one who has faith through activity and progress in a heaven here and not hereafter alone?" If he had spoken thus we venture to say there are thousands of labourers who would have understood him, just as they can understand Joseph Arch, because he touches the universal heart of toiling humanity, and any one who can make himself understood in that language, is sure to have a hearing among simple men.

Our commissioner concludes :—

"I will venture to say this much, that if electoral power should be entrusted to those only who are qualified by educational enlightenment to hold and exercise it, then for the present it will be more judicious to send John Whopstraw to an evening school than to make him a voter."

The best answer to this stale objection to enfranchisement of the unfitness of the labourers, is that of Frederick Douglass to the old objection to the emancipation of the slaves:—"The best preparation for freedom is freedom."

UNION BALLADS.

Singing has largely entered into the proceedings of the farm labourers' agitation. Most village meetings are begun with a song, and as the tune is generally something that, as Primitive Methodists, the majority of the men are more or less familiar with, and the words often a mere accommodation of some equally familiar hymn, the singing part of the meeting is not its least telling. A very favourite song is the following :—

STAND LIKE THE BRAVE.
(*Adapted by* G. M. BALL.)

O workmen, awake, for the strife is at hand ;
With right on your side, then with hope firmly stand—
To meet your oppressors, go, fearlessly go,
And stand like the brave, with your face to the foe.
    Stand like the brave, stand like the brave,
    Stand like the brave, with your face to the foe.

Whatever's the danger take heed and beware,
And turn not your back—for no armour is there ;
Seek righteous reward for your labour—then go
And stand like the brave, with your face to the foe.

     Stand, &c.

The cause of each other with vigour defend,
Be honest and true, and fight to the end ;
Where duty may lead you, go—fearlessly go,
And stand like the brave, with your face to the foe.

     Stand, &c.

We fight not alone who seek to be freed,
But friends from afar send us help when we need ;
And kindly they whisper, saying hopefully go,
And stand like the brave, with your face to the foe.

     Stand, &c.

Let hope then still cheer us ; though long be the strife,
More comforts shall come to the workman's home life :
More food for our children ; demand it—then go,
And stand like the brave, with your face to the foe.

     Stand, &c.

Press on, never doubting, redemption draws near—
Poor serfs shall arise from oppression and fear ;
Though great ones oppose you, they cannot o'erthrow,
If you stand like the brave, with your face to the foe.

     Stand, &c.

Another, which the witty secretary of the Union, Mr. Henry Taylor, composed on the occasion of our late illustrious visitor, the Shah of Persia's presence in England, takes well with the country folk. It runs thus :—

THE SHAH.

Tune—"Johnny comes Marching Home."

Pray who's this coming along the street?
        The Shah, the Shah.
He's jewelled from head to the soles of his feet ;
        The Shah, the Shah.
Now, let's go and meet him, and give him a greeting ;
Let's fuss him well up, and jolly well treat him ;
Such a noble fellow is
        The Shah, the Shah, the Shah.

A despot king of a serfdom land, is
        The Shah, the Shah.
Where his will is law, dealt out by the hand of
        The Shah, the Shah.
Where filth and pestilence, famine, decay,
Is of small concern, so his will they obey,
And robe with diamonds brilliant and gay,
        The Shah, the Shah, the Shah.

But he's come to learn a better way,
        The Shah, the Shah ;
He's come to learn what Christians say,
        The Shah, the Shah ;
How the arts and science of civilisation
Is the one great sign of a mighty nation,
And that despotic rule always means stagnation,
        Shah, Shah, Shah.

The glitter of fashion he don't want to see,
        The Shah, the Shah ;
He's sufficient of that in his own country,
        The Shah, the Shah ;
And if we *must* show him our cannon and spears,
And our vessels of war, of their strength when he hears,
No vain boast may he witness, let us tell it in tears, to
        The Shah, the Shah, the Shah.

THE REVOLT OF THE FIELD.

See our telegraph system and locomotion,
    Shah, Shah,
Our highway roads, and get a notion,
    Shah, Shah,
And strive to see, if ever should be
A famine again in your country,
By means of these, people fed may be,
    Shah, Shah, Shah.

See how our commerce in various ways,
    Shah, Shah,
The law of supply and demand obeys,
    Shah, Shah.
Free trade in all things we demand ;
In corn we have it, and soon the land
Will be made to serve the people's command,
    Shah, Shah, Shah.

Yes, Liberty's claimed to-day for all,
    Shah, Shah,
Serfdom and misery soon must fall,
    Shah, Shah.
Ev'n labourers, hitherto sadly neglected,
Have adopted a means by which they're protected
'Gainst tyrants and systems which made them dejected,
    Shah, Shah, Shah.

Th' National Agricultural Labourers' Union,
    Shah, Shah,
Is a means by which poor men in communion,
    Shah, Shah,
May strike off the fetters by which they're bound,
And help themselves and all who surround ;
'Twill never leave them where they were found,
    Shah, Shah, Shah.

So go back to Persia as soon as you please,
    Shah, Shah,
And instead of indulging and taking your ease,
    Shah, Shah,

Go tell the poor Persians, if they would be free,
In union they must all join heartily,
And remove all obstructions, e'en though it be thee,
      The Shah, the Shah, the Shah.

Mr. Howard Evans, of London, has composed a number of lively ballads, which have been published in a penny book, and are used very extensively at the out-door meetings. As the ballads of a people have always exerted a considerable influence over them, I give a few of these racy rhymes as illustrative of the men, their condition, the objects of their union, and the spirit which animates them.

THE FINE OLD ENGLISH LABOURER.

TUNE—"The Fine Old English Gentleman."

Come, lads, and listen to my song, a song of honest toil,
'Tis of the English labourer, the tiller of the soil;
I'll tell you how he used to fare, and all the ills he bore,
Till he stood up in his manhood, resolved to bear no more,
    This fine old English labourer, one of the present time.

He used to take whatever wage the farmer chose to pay,
And work as hard as any horse for eighteenpence a day;
Or if he grumbled at the nine, and dared to ask for ten,
The angry farmer cursed and swore, and sacked him there and then,
    This fine old English labourer, &c.

He used to tramp off to his work while town folk were a-bed,
With nothing in his belly but a slice or two of bread;
He dined upon potatoes, and he never dreamed of meat,
Except a lump of bacon fat sometimes by way of treat,
    This fine old English labourer, &c.

He used to find it hard enough to give his children food,
But sent them to the village school as often as he could ;
But though he knew that school was good, they must have bread and clothes,
So he had to send them to the fields to scare away the crows,
        This fine old English labourer, &c.

He used to walk along the fields and see his landlord's game
Devour his master's growing crops, and think it was a shame ;
But if the keeper found on him a rabbit or a wire,
He got it hot when brought before the parson and the squire,
        This fine old English labourer, &c.

But now he's wide awake enough and doing all he can,
At last, for honest labour's rights, he's fighting like a man ;
Since squires and landlords will not help, to help himself he'll try,
And if he doesn't get fair wage he'll know the reason why,
        This fine old English labourer, &c.

They used to treat him as they liked in the evil days of old,
They thought there was no power on earth to beat the power of gold ;
They used to threaten what they'd do whenever work was slack,
But now he laughs their threats to scorn with the Union at his back,
        This fine old English labourer, &c.

STAND BY THE UNION.

TUNE—" The Good Rhine Wine."

Stand by the Union ! all through the land
    The sons of the soil are waking ;
Join heart to heart, and hand to hand,
    The rusted chains of bondage breaking.
CHORUS—For the poor man is weak, though his cause be right,
    But the weak grow strong when they all unite.

Stand by the Union ! Labour's hope !
    One fibre is light as a feather ;

But the twisted strands of the good ship's rope,
  Defy the rage of wind and weather.
            For the poor man, &c.

Stand by the Union—the friend of all
  Who dare to befriend each other ;
Respond like men to the Union's call—
  He helps himself who helps his brother.
            For the poor man, &c.

Stand by the Union ! The great may frown ;
  We'll be their serfs no longer ;
Though they are strong who tread us down,
  The God-given rights of men are stronger.
            For the poor man, &c.

Stand by the Union, firm and true,
  We are bound to conquer through it !
We mean to win for toil its due,
  And we're the proper lads to do it.
            For the poor man, &c.

Stand by the Union,—onward we march
  For defence and not defiance ;
Our trusty chief is Joseph Arch,
  In right and union our reliance.
            For the poor man, &c.

Stand by the Union—stick to it now,
  With a strength no power can sever ;
We've put both our hands with a will to the plough,
  We'll never look back, boys, never—never !
            For the poor man, &c.

LORD REGINALD.

TUNE—"The Misletoe Bough."

Lord Reginald lives in a stately hall,
With flunkeys and keepers to come at his call,
With rich gems of art its walls are graced,
For his lordship has a most exquisite taste ;

On the same estate, as I happen to know,
Is a labourer's cottage, with one room below,
And one room above, where ten sleep in a batch,
And the rain pours in through the rotten thatch.

  Oh, the wrongs of the poor ! oh, the wrongs of the poor !

Lord Reginald lives on the fat of the land,
His wines are superb, his dinners are grand ;
And lately a prince to his house he brought,
For his covers afford such excellent sport.
But the people who live in the cottage, I ween,
Live not on the fat of the land, but the lean ;
For his lordship could hardly dine once at his club
On their weekly allowance for clothes and grub.

    Oh, the wrongs of the poor !

Lord Reginald's ancestors won by the sword
One half of their lands, and the rest by fraud ;
And now that the land by wrong has been won,
He has all the rights, and the people have none.
Lord Reginald's charities—trumpet them forth !
A few pounds a year spent in blankets and broth,
And allotments let at a very low rate—
The farmer pays two pounds, the labourer eight.

    Oh, the wrongs of the poor !

Lord Reginald does not his father's crimes ;
He speaks in the House and writes to the *Times ;*
Though his claws are sharp, his paws are sleek,
And he goes to the rich man's church every week ;
But with the pheasants his lordship kills
He pays his London fishmonger's bills ;
And when the peasants with hunger cry,
He prates of the laws of demand and supply.

    Oh, the wrongs of the poor !

Lord Reginald lives in the proud belief
That his family sprang from a Norman thief,
In the days when a robber was counted grand—
He conquer'd our fathers, and stole their land ;

But he kept them to till it, and gave them their share,
And the peasant still had enough and to spare.
Now I'd rather have lived under him, I think,
For *his* serfs had plenty to eat and to drink.
    Oh, the wrongs of the poor !

Lord Reginald wants to get more land still—
If he lacks the power, he don't lack the will ;
He'll steal our commons, if ever he can,
For he doesn't believe in the rights of man.
But the time is near when he'll have to be taught
That the land wasn't made to furnish him sport ;
That the down-trodden peasants, the children of toil,
By their labour and sweat have their rights in the soil.
    Oh, the wrongs of the poor !

FARMER GRUMPS.

Tune—" Poor Mary Ann."

I'll sing you a song of our mutual friend, sir,
    Poor Farmer Grumps ;
He don't sing himself, you may depend, sir,
    Poor Farmer Grumps ;
For care and trouble never leave him,
All things were made to vex and grieve him,
All men to worry and deceive him,
    Poor Farmer Grumps.

He grumbles on for weeks together,
    Poor Farmer Grumps ;
He grumbles when it's rainy weather,
    Poor Farmer Grumps ;
He grumbles when the sky is sunny ;
And yet, to me, it's rather funny,
Somehow or other he still makes money,
    Poor Farmer Grumps.

Whenever they call on him for taxes,
    Poor Farmer Grumps,
Into a passion at once he waxes,
    Poor Farmer Grumps ;

But while he thus the Poor Law curses,
He thinks not how their case far worse is
Who have to trust to its tender mercies,
    Poor Farmer Grumps.

He growls to you like a sore-headed Bruin,
    Poor Farmer Grumps,
That his landlord's game will be his ruin,
    Poor Farmer Grumps.
But do not think that his defection
Will add one vote to the Liberal section;—
He'll crawl to the squire at the next election,
    Poor Farmer Grumps.

And now he's in a frightful rage,
    Poor Farmer Grumps.
For his men have asked for a larger wage,
    Poor Farmer Grumps.
So, like a fool, at once he sacked 'em;
But he soon found out that the Union backed 'em,
And off to the North to work it packed 'em,
    Poor Farmer Grumps.

He knows he's made an awful blunder,
    Poor Farmer Grumps.
Where will he get fresh men, I wonder?
    Poor Farmer Grumps.
His cares have made him so much thinner,
And as the Union proves the winner,
Let us forgive the repentant sinner,
    Poor Farmer Grumps.

THE MASTER AND I.

Tune—"Bonnie Dundee."

Says the master to me,—" Is it true, as I'm told,
Your name's on the books of the Union enrolled?
I can never allow that a workman of mine
With wicked disturbers of peace should combine;

I give you fair warning, mind what you're about,
I shall put my foot down on it, trample it out ;
On which side your bread's butter'd sure you can see,
So decide now at once for the Union or me."

Says I to the master,—" It's perfectly true
That I'm in the Union,—I'll stick to it, too!
And if between Union and you I must choose,
Why, I've plenty to win and little to lose ;
For twenty years mostly my bread has been dry,
And to butter it now I shall certainly try ;
And though I respect you, remember I'm free—
No master in England shall trample on me."

Says the master to me,—" A word or two more—
We never have quarrelled on matters before ;
If you stick to the Union, 'ere long, I'll be bound,
You'll come and ask me for more wages all round.
Now I cannot afford more than two 'bob' a day,
When I look at the taxes and rent that I pay ;
And the crops are so injured by game, as you see,
If it is hard for you, it's hard also for me."

Says I to the master,—" I do not see how
Any need has arisen for quarrelling now ;
And though, likely enough, we shall ask for more wage,
I can promise you we shan't get first in a rage ;
Your landlord mayn't think so, but I don't deny
Your taxes and rent are a great deal too high ;
But granted all that,—you've been both in the swim,
We are bound to squeeze you,—now you go and squeeze him."

THE GREAT UNPAID.

TUNE—" Sammy Slap, the Bill Sticker."

My name it is Squire Puddinghead,
   A Justice of the Peace, sir;
And if you don't know what that means,
   Just ask the rural police, sir !

When culprits nabbed for petty crimes
   Within my Court assemble,
If I am sitting on the Bench,
   Oh, don't the wretches tremble
           At the Great Unpaid !
           Ask anything but justice
           Of the Great Unpaid.

The cases that I have to try
   Are mostly small transgressions—
So small, the Court in which I sit
   Is called the Petty Sessions—
A sort of legal small-tooth comb;
   Th' offences are so tiny,
You'd laugh at them,—but you'd not laugh
   When I proceed to fine ye.
           Oh, the Great Unpaid, &c.

If Polly Brown but takes a stick
   From Farmer Giles's fences,
I fine her twopence as its worth,
   And fourteen bob expenses;
And if a tramp sleep in a field,
   Such is my lordly bounty,
I give him lodging for a week,
   Provided by the county.
           Oh, the Great Unpaid, &c.

The Union leaders I would hang,
   'Twould be a task delightful:
But since I can't, I am content
   To do the mean and spiteful;
And if my colleague, Captain Fair,
   Would be the poor's protector,
The vilest things I dare to do
   Are backed up by the Rector.
           Oh, the Great Unpaid, &c.

So Policeman Hobbs, and Snobs my clerk,
  Their paltry charges trump up,
To vex and harass Union men,
  And don't I make 'em stump up!
What good to me to be J. P.
  Over my wretched drudges,
If I can't strain and twist the law,
  To pay off all the grudges
          Of the Great Unpaid, &c.

THE TWO UNIONS.

TUNE—"The Red, White, and Blue."

Who would ever have thought we'd be making
  The Union a subject of song,
When our hearts with distress were nigh breaking,
  And there seemed no redress for our wrong.
But the Union I mean is another,
  A Union of good men and true,
That in time shall abolish the other,
  For the old must give place to the new.

      CHORUS—For the old must give place to the new,
             For the old must give place to the new,
             The Union of Labour for ever
             The old must give place to the new.

Once the name of the Union was hated—
  That Union that's worse than a jail—
Where husband and wife are unmated,
  When age makes them feeble and frail.
But, much as we hated and feared it,
  We had no better prospect in view,
Till Arch wrote on his flag, and upreared it,
  "The old shall give place to the new."
          For the old, &c.

Good wage is the great curse of labour,
  His lordship in confidence prates.
Let him give what is just to his neighbour,
  And he'll get back the cost in the rates.

Mark! farmers and squires, 'ere I finish,
  If you give but to labour its due,
Your Unions will greatly diminish,
  For the old will give place to the new.
                      For the old, &c.

O, THE ROAST BEEF OF OLD ENGLAND!

NEW VERSION.

I'll sing you a song—it shall not be too long,
If you go for your rights, you will not think it wrong,
So give us a chorus, both hearty and strong;
      O the prime roast beef of Old England,
      O the rare Old English roast beef!

Time was when our ancestors farmed their own land;
Contented and happy, no foe could withstand
Those bold sturdy yeomen—a glorious band,
      Well fed with the beef of Old England,
      The jolly Old English roast beef!

But, somehow or other, the lord and the squire
Have got all the lands that their hearts can desire,
And we are poor wretches just let out for hire,
      And have lost the roast beef of Old England;
      We don't get a smell of the beef.

A mite of fat bacon, a bit of hard bread,
Our wages will much about give us instead,
Yet we think those that work have a right to be fed,—
      Aye, even with beef of Old England,
      With jolly Old English roast beef.

The lord and the squire may open their eyes,
The parson's next sermon be wonderful wise,
We hope their fine feelings we shall not surprise,
      But we want the roast beef of Old England,
      The jolly Old English roast beef.

So pull altogether, and work with a will,
For something our wives and our children to fill,
Much better than crusts, and their tea-kettle swill,
    For jolly roast beef of Old England,
    For glorious Old English roast beef.

So stand by the Union, the winter's gone through,
Neither hunger nor cold could our courage subdue,
For there's one thing we want, and mean having it, too,
    The jolly roast beef of Old England,
    The glorious Old English roast beef.
BENJAMIN BRITTEN.

A HYMN OF LABOUR.

LORD, as thou didst invite of old,
    We weary, heavy-laden come;
Thy roused, uprising poor behold,
    And those with hopeless misery dumb.

Beneath our feet we tread the lie,
    That our great wrongs are Thy design,
That we in want should live and die,
    While others share the corn and wine.

Crowned Justice! hear our cry of wrong:
    Throned Love! attend our wail of pain;
Plead Thou our cause against the strong,
    Who serve as god, the Moloch gain.

Thy fair sweet-smiling fields they stain
    With many a foul and shameful blot;
Rises to heaven the cry of Cain—
    "My brother's keeper I am not."

We thank Thee for the men who lead,
    Who fight our cause with tongue and pen;
Whose love to Thee,—best shown in deed,
    Breaks forth in ardent love to men.

We thank Thee, that from north to south,
    From east to west the flame has spread ;
And that the breathing of Thy mouth,
    Has kindled into life the dead.

Lord, make us patient, as Thou art,
    Yet constant to our great design ;
From thoughts of vengeance keep each heart,—
    Justice and love are both divine.

More men, more manhood now accord,
    Make us more worthy to be free ;
Where dwells the spirit of the Lord,
    There is the home of liberty.

THE FRANCHISE.

TUNE—"The Englishman."

There's a man who represents our shire
    In the Parliament House, they say,
Returned by the votes of farmer and squire,
    And others who bear the sway ;
And farmer and squire, when laws are made,
    Are pretty well cared for thus ;
But the County Member, I'm much afraid,
    Has but little care for us.
CHORUS—So we ought to vote, deny it who can,
        'Tis the right of an honest Englishman.

Whenever a tyrant country beak
    Has got us beneath his thumb,
For Justice then he ought sure to speak,
    But the County Member is dumb.
Whenever the rights of labour need
    A vote on a certain day,
The County Member is sure to plead
    And vote the contrary way.
          So we ought to vote, &c.

We ask for the vote, and we have good cause
  To make it our firm demand;
For ages the rich have made all the laws,
  And have robbed the poor of their land.
The Parliament men false weights have made,
  So that Justice often fails;
And, to make it worse, "The Great Unpaid"
  Most always fiddle the scales.

    So we ought to vote, &c.

The subject of the last song has furnished our facetious friend "Punch" with a topic for his graphic pencil, and his clever hit drew forth the following rejoinder, which I take from the columns of the "Chronicle."

GAARGE TO "PUNCH."

"Punch" last week had a cartoon representing Mr. Gladstone giving Hodge a vote, with the following words printed beneath it :—

"HODGE.—Lor a massy measter, be oi 'a power in t' staate!'
  What be oi to get by that?
Mr. G.—That my good friend is a mere matter of detail.
  The question is, what am I to get by it?"

MA-ASTER PUNCH,—
Oi seed yewre pictur in a shop laast market day,
Vor varmer Grumps sent oi to tawn along wi' a loa-ad of hay,
An' a oastler chap oi knows there as a got a lot o' cheek,
Says they've bin an' put yeure loikness into "Punch" this week.
Well oi looks at un a bit ; says oi—"What be un all about?
Oi dunno what un mea-ans, oim blowed if oi can ma-ak un out;"
So a sma-artish lookin chap theere as had just turned round to gow,
Says, "Him as is scratching on his yead, is mea-ant for yew yew
 know."

Says oi, "If that be what un mea-ans, the pictur's very poor,
Oi never seed a labourer as looked loike he avoor ;
T' chap as ma-ade that pictur was i' Lunnon born and bred,
It's loike the picturs ower Bill dra-as all out o' his own yead,
An' when he dra-as un on t' slate he has to write below un :
' This be a mon,' ' this be a house,' or nawone wouldn't know un."
An' shewer enough *yew'd* put a lot o' writin onderneath
To tell us what that chap was mea-ant vor, wi' the grinnin teeth.
Well in oi gaws ; says oi to t' gell, "What be that pictur theere?"
"Thruppence," says she ; "oi bought un, but oi thowt un moighty
    dear."
Why Ma-aster Punch our eddytoor gives ten times more'n yew,
Ower pa-aper's three times less in price, and better reading too,
An' as for t' picturs, if that one yew give about the voat
Is t' best that yew can do, whoi then I say they're good for nowt.
Well, soon as I gets whoam and puts t' oarses op all right,
Oi gaws down to t' Union Branch, for t' wor ower meetin night ;
Says oi, and pulls t' pictur out, an' shows t' chaps all round—
"Old Punch can't find a live un loike him I bet a pound."
Ower Sekketary then gets oop and reeads the writin to the men ;
Says he, " Punch thinks we all be fools because he be hissen."
We talked un over more'n an hower, and so at last to end un,
They passed a resolution and said oi woor to send un.
" Resolved—that we, the labourers of —— are not so green
As surten stoopid people think, for we know what we mean.
We beant Whigs nor Tories, an' we doont keer which be in,
Vor Gla-adston and Disraali we wouldn't give a pin ;
If a Toory cooms to ask us if we'll give un a voat,
We're shewer tell un naw becaas we know *he's* good for nowt ;
And if a Libreal cooms that be a loard's or bishop's broother,
We know it's ten to wun but what he's joost as bad as t'oother.
If he shouts Gla-adston, we shall ask un—how abeaut t' gaame ?
If he beant right on that, we'll send un back from where he caam.
We'll ask un if he mea-ans to gie us all the wa-aste lands back ;
Git joostice done the poor, and give the great onpaid t' sack,
Will un give back all t' church lands to t' staate from wheer it got 'em,
An' leave t' choorch like oother sects to stand on its own bottom :
If he says naw, I woant, we'll tell un to be goan,
An' put in soome rough spoaken chap wi' a foostian jacket on."

Oi daare say yew will find this boan be moighty hard to crunch,
But shewer as oim I livin mon we'll do it, Ma-aster Punch.
By this heere resolution you will see ower platforms laarge,
But we'll find a mon to voat vor 't as shewer as my naam's
GAARGE.

A sort of pet-project of Mr. Arch's is the securing to labourers three or four acres of land at an average rent. He sees in that the only hope of permanently satisfying the men. Wherever it has been fairly tried, he says the result has been satisfactory. The bane of utter dependency on the employer is removed, and with it the badge of pauperism.

Among what may be called the literature of the movement, I find the following apparent confirmation of the correctness of Mr. Arch's calculations in the matter. The extract is long, but as the subject is of growing interest just now I prefer inserting it whole. The writer, as will be seen, is replying to an article on the labourers' question in "Fraser."

THE LABOURER AND THE LAND.

In many cases it appears that knowledge does not progress. The idea of the widening of the thoughts of men by the process of the suns seems mere poetry or vanity if we test it by evidence common enough. Every generation seems to have started afresh in the knowledge of some common things; for instance, "A Wykehamist" in "Fraser" for June, speaks of the labourer and the land as if he had never read or never thought of the return the soil gives to spade cultivators; and yet without the least knowledge of the subject he ventures to write for educated readers his views of the profit of the cultivation of the soil by the peasant. If this "Wykehamist" represents the average information of the educated and higher classes, the ignorance is so serious in its consequences that we must diligently remove it. If the aristocracy get the silly idea generally diffused amongst them, that twenty acres of land will not return to the peasant for his labour half a living, it may seriously injure the prospects of the peasantry being restored to the land.

Few of our peasantry would refuse the chance of getting a living off five acres of good land, moderately rented; and the majority would believe their fortunes made if they had ten acres of good land on fair terms. And certainly their judgment is much wiser than the county member who writes under the *nom de plume* of a " A Wykehamist" in " Fraser."

Surely our villagers will laugh at the absurdity of a public writer teaching as follows :—

"If an arable farmer in prosperous times makes £1 an acre profit, he will be a very lucky being. How, then, shall the cottier without any of the advantages of machinery make £20 a year off his twenty acres? And £20 a year would be considerably below what his day labour would bring him in."

It would be a very difficult task to pen a more silly statement about the land and the labourer than this. A spade is a machine, and the cheapest and most effective in willing hands in working the soil.

However, we shall proceed to give facts, and those of long standing, from established authorities, such as the "Quarterly Review," Vol. xli., 1829.

Sir Henry Vavasour, writing to Lord Carrington, then president of the Board of Agriculture, says :—

"I have had the honour of mentioning to your lordship the advantages that appeared to me in cultivating land in the Flemish manner, or what is now called about Fulham and that neighbourhood, the field-gardening husbandry. I have for some years encouraged my cottagers in Yorkshire in this mode of managing their small garths, or gardens, which are in general from one to three acres; and I have now an opportunity of stating the husbandry of a poor industrious cottager's garth. The man's name is Thomas Rook; as he can neither read nor write, these particulars have been transmitted to me from his own mouth; and as I saw his land almost every day during the last harvest, I can vouch that this account is not far from the truth.

Quantity of land.

A.	R.	P.	PRODUCE.	VALUE.
				£ s. d.
0	2	0	240 bushels of potatoes	24 0 0
0	1	0	60 do. of carrots	6 0 0
0	3	20	5 quarters of oats at 44s.	11 0 0
1	0	0	4 loads of clover, part in hay, part cut green	12 0 0
0	0	20	Turnips	1 0 0
0	0	30	Garden stuff for family—peas, beans, cabbages, &c.	
3	0	0	Gross produce	54 0 0
		Deduct	{ Rent of cottage and land 9 0 0	
			Seeds 3 0 0	
			Labour 10 10 0	
				22 10 0

Profit from the cultivation of three acres, exclusive of butter and the garden stuff consumed in the family 31 10 0

His stock was two cows and two pigs; one of his cows had a summer's gait for twenty weeks with his landlord. The land was partly ploughed and partly dug with a spade, cultivated (the ploughing excepted) by the man, his wife, and a girl about twelve years of age, in their spare hours from their daily hired work, seldom a whole day off, except in harvest; made the rent in butter, besides a little used in the family. The man relates that he thinks he clears, one year with another, from the three acres, about £30. The daily wages his family earn about keep them. It is very evident that this man clears from his three acres more than a farmer can possibly lay by from more than eighty acres of land in the common husbandry of the country, paying for horses, servants, &c.; and it must be obvious to every one how great the advantages must be to society by cultivating land in this manner. It would have taken more than half the quantity of his three acres in pasture for one cow at grass during half the year; whereas (excepting the summer's gait for one of his cows, as mentioned before) his stock of two cows and two pigs is kept and carried on the whole year; the family lives well, and a handsome sum has been yearly saved to place out two sons and supply them with clothes and other necessaries."

The notion of "A Wykehamist" that he thought valuable enough to print in "Fraser," is that a peasant must be a clever fellow to get £20 off twenty acres, while the real experience of a peasant is that he can get £31 or more off three acres! And this fact was attested by the late Sir Henry Vavasour. Surely "A Wykehamist," who is, if we are rightly informed, a county member, must be heartily ashamed of himself to print statements utterly without foundation in truth, many exceedingly improbable, and if he has a fact to quote he invariably mars it with blundering.

The statement of Sir Henry Vavasour influenced a gentleman, Charles Howard, Esq., who had bought some poor land at £10 an acre, to let a cottager have one acre and a quarter of the most barren. And when the man had got it in order, and covered with crops, his landlord measured and valued the produce, and found it amounted to ten pounds seventeen shillings, and yet he valued every item at less than the market price; just about the price of the fee simple of the land, raised by a peasant by the crops of one year!

We quote another instance from the same authority:— "A labourer at Hasketon, in the county of Suffolk occupied four inclosures, containing fourteen acres of pasture land, at a rate of £13 per annum, upon which he kept two cows. He died, and these two cows, with a very little furniture and clothing, were all the property that devolved, upon his death, to his widow and fourteen children, the eldest being a girl under fourteen years of age. The parish officers, upon being made acquainted with the situation of the family, immediately agreed to relieve the widow by taking her seven youngest children into the house. This was proposed to her, but with great agitation of mind she refused to part with any of her children. She said she would rather die in working to maintain them, or go herself with all of them into the house, and work for them there, than either part with them all or suffer any partiality to be shown to any of them. She then declared that if her landlord would continue her in the farm, as she called it, she would undertake to maintain and bring up all her fourteen children, without parochial assistance. She persisted in her resolution, and being a strong woman, about forty-five years old, her landlord told her she should continue his tenant, and hold it the first year rent free. This she accepted with much thankfulness, and assured him that she would manage for her family without any other assistance. At the same time, though without her knowledge, Mr. Way, the landlord, directed his steward not to call upon her at all for his rent, conceiving it would be a great thing if she could support so large a family even with that advantage. The result, however, was, that with the benefit of her two cows and of the land, she exerted herself so as to bring up all her children, twelve of whom she placed out in service; continuing to pay her rent regularly of her own accord every year after the first. She carried part of the milk of her two cows, together with the cream and butter, every day to sell at Woodbridge, a market town two miles off, and brought back bread and other necessaries, with which, and with her skim-milk, butter-milk, &c., she supported her family. The eldest girl took care of the rest while the mother was gone to Woodbridge; and by degrees, as they grew up, the children went into the service of the neighbouring farmers. She came at length and informed her landlord that all her children except the two youngest were able to get their own living, and that she had taken to the employment of a nurse, which was a less laborious situation, and at the same time would enable her to provide for the two remaining children, who, indeed, could now almost maintain themselves. She therefore gave up the land, expressing great gratitude for the enjoyment of it, which had afforded her the means of supporting her family under a calamity which must otherwise have driven both her and her children into a workhouse."

What a heroine this woman was! She deserved a statue far better than any prince or peer we have ever known. A woman kept, and fourteen children reared and provided for, out of the profit of two cows and fourteen acres of land! And "A Wykehamist," probably an M.P. for a county, actually proclaims to the world, through "Fraser's Magazine," that a cottier "cannot make £20 a year off twenty acres." Imbecility on the part of "A Wykehamist," in contrast with the quick comprehension and noble industry of a peasant's widow! And the management of the land was not at all equal to what can be done now. Some of these days we shall be explaining how a cow can be kept upon a single acre of land, or three upon two acres; and Cobbett taught that a cow could be kept on much less than an acre; and a Mr. Thornton kept three cows upon seven roods of land, poor and sandy, near Huddersfield.

Triumphant as we have been with our evidence over the

statements of "A Wykehamist," we are disposed to carry the proof still further—produce an instance of farmers ruining a parish, leaving it a hopeless place, the prey of pauperism, and the paupers raising the parish to prosperity again, and more, becoming contributors to another parish, assisting farmers to support the pauperism they had made, and yet could not maintain :—

"The parish of Cholesbury, in Buckinghamshire," says Sidney Smith, "was entirely occupied by two large farmers. Fertile, populous, within forty miles of the metropolis, its cultivators notwithstanding fell behind. There were one hundred and thirty-nine inhabitants in the parish, but only two had an inch of the soil. Was not this civilisation run mad? Was it not a glaring and staring evidence of the monstrous abuse of the principle of private property, that only one man out of sixty-nine tillers of the ground should have exclusive occupation of the earth, which God made common to all, and the appropriation of which can only be palliated upon the clearest proof of public advantage? What was the consequence of this *beau ideal* of politic-economical arrangement? Simply this: out of the one hundred and thirty-nine inhabitants, one hundred and nineteen were paupers. The land monopolists became bankrupt, the parson got no tithes, the landlord's acres were in rapid course of being eaten up with rates, and the whole property of the parish being unable to feed the inhabitants, a rate-in-aid had to be levied on the neighbouring parishes, which were rapidly degenerating into the same state. The Labourers' Friend Society came to the rescue. They leased the land at a fair rent. They parcelled it out among the very worst class of persons upon whose habits to hazard the result of such an experiment. Some got five, some ten acres, according to the size of their families, and what was the effect? At the end of four years the number of paupers had diminished from one hundred and nineteen to five, and these were persons disabled from old age or disease—these paupers afforded to pay a rate-in-aid to the neighbouring parishes; and it was found that every one of them were in a state of independence and comfort, each had a cow, many two or three, to which some added a horse, others some oxen ready for the market, and all had pigs and poultry in abundance. No experiment could be more severe than this. Persons once degraded to the condition of paupers lose self-respect, the love of independence, the spirit of self-help."

No sketch of this strange rural awakening would be complete which did not notice the attitude of hostility to the Established Church of the country which its leaders have from the first assumed. The fact of their being for the most part Primitive Methodists, explains it in measure, but other causes must be looked for, in order to a full explanation of the somewhat startling phenomenon. Mr. Goldwin Smith has, I think, furnished a clue to the mystery—a mystery when we remember the eloquent pleadings of the present Lord Chancellor in the House of Commons a few years ago, on behalf of the Establishment, on account of its assumed advantages to the labourers in our rural districts. Mr. Smith, in a speech at the Trades' Union Congress at Sheffield, did not deny that the parish clergyman dispensed a good deal of charity in the villages, but he argued that the kindness is usually such as tends to keep a man *contented with a more or less wretched position*, and not that more enlightened and truer kindness which helps him to help himself. There are many exceptions, but the rule has been, that there is nothing which squires and parsons as a whole have striven less to do, than to make the labourers *permanently independent of their bounty*.

The following extracts from an article in the "Labourers' Chronicle" fully sustain this explanation of the circumstance. It is one of a series of strongly-worded anti-state Church articles, entitled, "The Church of England in its Relation to the Labouring Classes."

"The Church of England," says the writer, "is within a few centuries as old as Christianity. It has grown with our country's growth, and strengthened with her strength. Its foundations lie in every parish and hamlet in the land. Its bishops sit in Parliament, its clergy dominate society; the wealthy classes everywhere pay them homage. The Church is an undoubted power in the land—a power too great to despise, too palpable to ignore; but in proportion to its power is the greatness of its responsibility. Did it do the true work of a Church it would be the greatest of national blessings, for it would then bear to all ranks and conditions of men a witness for the things that are true, and noble, and good, and it would so mould public opinion, that no great social wrong could live beneath it. We say, first, that the Church of England is one of the most powerful institutions in the country; secondly, that any social, or moral, or religious reform which this Church saw and felt to be necessary, it has, if it chooses to use it, full power to effect. Had it commiserated the wretched condition of hopeless and ill-paid toil, of life-long poverty, and social, intellectual, and moral degradation in which the peasantry of England have for years past been living, while their employers were growing rich and living luxuriously on their labours; had the Church done a Church's duty in this matter, it would have raised such a protest against the monstrous injustice, against the oppression and spoliation of the weak and helpless poor; would have filled the minds of the middle and upper classes with such a sense of burning shame at the pitiful wages paid by those who held the land to those who cultivated it; at the monstrous robbery and wrong which defrauded the labourer of the rightful reward of his hard and honest toil, that justice would long since have been done, and a happy peasantry, owning the Church as their best and truest friend, and shaping their lives by its wise and loving counsels, would have filled the places now occupied by miserable, and starving, and discontented serfs."

Failing here, the writer pronounces the Church a failure everywhere.

"Such a Church," he continues, having dwelt upon an assumed absence of humanitarianism in her ordinary teachings, "teaches men to serve God without helping their brother, and deludes them by the hope of winning heaven hereafter, while they lead selfish and ignoble lives here. We are no opponents of religion; we are its friends, not its enemies. To a true Church of God we would give all possible reverence. We need the word of life which such a Church would speak; the moral guidance such a Church would give; the ennobling influence which such a Church would diffuse. We need the summons to a real manhood such a Church would ever speak; the rebukes it would give to the pride of the rich; the encouragement with which it would gladden and strengthen the strivings of the poor; the reproof it would send to the vices and follies of all. We would have the

Church of England to be such a Church as this, and to take up this work, for till this is its character it has no claim to be a Christian Church at all, much less to be supported by a free people as an institution of the land. We say emphatically that the Church which does not labour to bring temporal blessings to the poor, neither reads the Bible correctly, nor interprets Christ's ministry aright.

"Apart, however, from the sanctions which the State confers upon the Church as a national institution, of which the Queen is the head, the Church of England is just now asserting a corporate independence, claims to hold and exercise a higher authority than that which the State gives—an authority which is spiritual, not temporal, in its character. The claim is a valid one. He who speaks the truths of religion, whether he be priest or layman, speaks in the name of God, and should speak the word and the truth of God—that word of life which all should reverence and obey. In the world's history, however, this assertion of spiritual authority has been the subject of greater abuse than has the assertion of temporal authority, and priestcraft has ever been a worse oppression than kingcraft."

This latter paragraph reveals another great occasion of stumbling, as regards the labourers and the Church, and fully accounts for all that remains to be accounted for, of the well-nigh universal hostility of this movement to it as an establishment. In most country villages there is a development of priestism which presses hard upon the consciences of Dissenters. Their children must either remain in ignorance, or be subjected to teaching and influences which they abhor. Parish gifts and charities are extensively utilised as aids to devotion by the new order of priests. To persist in going to the village chapel, the labourer has found to mean the loss of the five hundredweight of coal at Christmas, while the consequences to his children of his stern determination to have them on the Sunday at the Chapel instead of the Church school, are sundry pains and disabilities which make them social pariahs in their respective neighbourhoods. I have found this sort of petty persecution more common in our rural districts than I was at all aware of, and doubtless it helps to elucidate the phenomenon now under consideration. The article concludes thus :—

In its early days the Church of England was wholly and entirely an institution of the State, and there was no essential difference between clerics and laymen. Before the time of William the Conqueror, laymen sat and voted in Church Synods, and the clergy were members with the laity of the Wittemagemot, or Parliament. In this reign, however, the machinations of Rome succeeded in effecting the introduction of canon or clerical law, and withdrawing the clergy largely from civil jurisdiction, virtually setting up the Church as a power concurrent with the State, and in spiritual matters superior to the State. Through the succeeding reigns there was a constant battle between Church and State —that is, between the Pope and bishops on the one hand, and the king and Parliament on the other hand. In King John's day, the Pope nearly attained a complete supremacy. This led, however, to a manifesto on the part of the barons, and the declaration of Magna Charta—the great assertion of the civil rights and liberties of the people. In Henry VIII.'s reign, the Pope and his party sustained a thorough and crushing defeat, and the religion of England was reformed in doctrine, and wrested from the Pope's dominion. Still the rags and tatters of Popery remained, and the remnants of this evil system may be seen to-day, not only in the sanction which the Prayer-Book gives to many Popish rites, but in the claim which the Church makes to be a spiritual corporation above the interference and the control of the State, having the right to govern the people through Convocation in matters of religion, while Parliament legislates for their temporal affairs; and as a result of this assertion, we find the Church almost entirely confining itself to what it calls the spiritual welfare of the country, and ignoring all responsibility as to aiding social reforms. Hence the creeds and dogmas which relate so largely to the soul's welfare in the next world, and which do no little towards making this life more tolerable to the toiling masses of the people.

We assert that the Church of England is a national institution; that in a country which possesses household suffrage, it should be held to its proper work of bearing witness for truth, and right and justice, and all things noble and good; that it should be the friend, and helper, and protector of the poor, and set before the minds of all men the ideal of a true and noble manhood. It is an institution that exists for the popular welfare; that receives its discipline from the laws of the country, but that should find its inspiration from the noblest promptings of human souls, and the wisest utterances of human minds.

The Church of England is the national instructor of the people in the things that pertain to character and conduct in the principles and philosophy of life, in all that relates to the righteousness that exalts a nation. Its business is to put down all evil, to reprove all rampant injustice, to demand a remedy for all social wrong. It is, under Providence, the appointed guardian of the poor, the helpless, and the weak; it should be the great peacemaker in our midst, the guide and friend of all. We ask, however, have the labouring classes cause to thank the Church for service such as this? Did "the tidings of great joy," which spoke of better wages and happier lives, which told the serf he was a man, and claimed for the toiler a right to live in decency and comfort, reach him from the national pulpit? No. Five hundred thousand voices will emphasise the answer, and declare that the Church of England has never extended to them even the shadow of such service as this. Clergymen of this Church who live in luxury upon its revenues; deans, and canons, and bishops, who rival princes in the splendour of your palaces, and to whom the cross is now but the symbol of those temporal glories that the devil once vainly pressed on your master's acceptance, remember that the people are in power to-day, and they are about to test you by your work, by the value of the services you render to themselves. They need a real and true Church, and they are prepared to love, and reverence, and support a Christian clergy who will do their Master's work. They want none of your cursing creeds —none of your pretended sacraments; they want counsel and guidance for good and true living, and will value you only as you prove to them to be a helper and a friend. Already they are weighing you in the balances. Be wise, then, and be warned in time; or, when the pent-up storm breaks loose, you and your religion will be rudely swept away, and the places you have so unworthily occupied will be given to better, and truer, and more Christian men to fill.

The writer of this somewhat strongly-worded article, gives what I have no doubt is the more or less uttered or unexpressed verdict of the executive of the movement. Rightly or wrongly, the labourers have looked to the Church for sympathy in their endeavours to throw off their semi-serfdom, and they have looked in vain. As I have before remarked, when their leaders appeared on the platform of Exeter Hall, to appeal to the country, the only clergyman who ventured to come to their support was Archbishop Manning. In multitudes of instances there has been superadded to this sin—as they have regarded it—of omission, sins of commission. Their speakers in villages and rural towns have been denounced as violent agitators—firebrands, communists, and other opprobrious names.

A most unfortunate occurrence at Chipping Norton, last year, greatly intensified this feeling on the part of the labourers against the Church. Although the facts of the case are tolerably well known to every one, it may be as well, perhaps, to incorporate them into my narrative. The first intimation of the affair that reached the eye of the public was the following letter which appeared one May morning in the "Times." The writer is one of the members of the executive committee of the Labourers' Union, as well as a local official, as stated in his letter.

IMPOSSIBLE!
COMMITTAL TO PRISON OF SIXTEEN WOMEN AT CHIPPING NORTON.

SIR,—Please use the following facts, by way of publication, in your valuable paper:—I am Chairman of the Oxford district of National Agricultural Labourers' Union, and at Ascot, Oxfordshire, in my district, there has been, for more than three weeks, a lock-out of the farm labourers, and considerable excitement has prevailed in consequence. Mr. Hambridge, a farmer, has brought men from other villages to do the work of his former men, and several of the women of the village, whose husbands are out of work, assembled on the 12th instant, and meeting Mr. Hambridge's men as they were going to work, tried to induce them to leave their employment. Mr. Hambridge took the matter up, and caused seventeen of the women to appear to-day before the magistrates at Chipping Norton, in this county, and the magistrates have sent sixteen of these poor women to prison—seven of them for ten days' hard labour, and nine for seven days' hard labour. The witnesses swore that some of the women had sticks in their hands, and told the men that they would not be allowed to go to work. The women denied this altogether. Anyhow, it was not attempted to prove that any physical force had been made use of. The women are very respectable in the class to which they belong. There was great excitement at the town of Chipping Norton on the decision of the magistrates being made known, and a crowd soon assembled, and it was with great difficulty that I could restrain their anger. I fear that if I had not been present some violence would have been committed. The people were astonished and bewildered at the sentences on this number of poor women, for what appeared to them no offence whatever. The women themselves had no idea that such a law existed as was now brought to bear on them with such terrible force. I attended the meeting in the capacity I hold, to watch the case, and I was never more shocked than to see these sixteen poor women dragged off to prison, and some with infants at their breasts. The magistrates present were—Rev. Carter, of Sarsden, Rev. Harris, of Swarford.

C. HOLLOWAY.

COMMITTAL OF WOMEN TO PRISON. 133

The immediate result of this letter was the despatch of a special commissioner from the "Times'" office to Chipping Norton. The heading of the letter, with the significant word, "Impossible," boded ill for the parties concerned in the lamentable proceeding. The following report of the case was published the next day in the leading journal:—

Very considerable excitement at present prevails throughout a large district of Oxfordshire, in consequence of the committal to prison of sixteen women by the Chipping Norton Bench of Magistrates on Wednesday last. In a letter which appeared in "The Times" of the 23rd inst., Mr. C. Holloway, of the National Agricultural Labourers' Union, gave a substantially accurate account of the circumstances under which the justices dealt with these women, but falling into an error not uncommon with unionists, Mr. Holloway called, what is in reality a "strike" on the part of certain agricultural labourers, a "lock-out" on the part of the masters. It is but fair to him to say that on this being pointed out to him, he said he had supposed the two terms were synonymous. The facts of the matters in dispute between masters and men, and the steps taken by both sides in consequence of those differences, he states with complete frankness, and there seems to be no reason to doubt that his version of those matters is perfectly truthful. The propriety of the course taken by the organisation, of which he is a district chairman, is quite another question.

That the public may rightly understand the facts connected with the decision at which the Chipping Norton magistrates arrived, it would be well to go back a little, and state how it was that dissensions between the employers and employed grew up, in a neighbourhood remarkable for its generally pacific character, until two magistrates adopted a course which, to say the least of it, was extremely harsh and singularly ill-advised, and which has led to serious riot and wide-spread uneasiness. Into the origin of the movement of which Mr. Arch is the apostle, it is not necessary to re-enter. The public are fully informed on that subject; but it may not be so well known that since Mr. Arch set that movement on foot the Union has made strong, and not altogether unsuccessful efforts to establish itself in Oxfordshire. It is gratifying to be able to state, on the evidence of Union leaders, that several of the landed proprietors in that county have all along treated their labourers with greater liberality than even the Union demands. It asks, and backs the men in requiring average wages of 14s. a week. Various gentlemen have been paying more than that, but the Unionists say that when they began operations they found the average wages paid by the farmers only 11s. They boast that they have brought this up to from 12s. to 14s., and their plan has been to direct partial strikes when the masters refuse to concede the terms the men were authorised by the Union to demand. The headquarters of the Union, for this part of the country at least, is Leamington, and there are two or more districts in each of the counties administered from those headquarters. One of the districts for Oxfordshire has its local commanders at Woodstock.

The levy off each labourer in employment is 2d. per week, and that there must be very numerous adherents to the Union in the district is shown by the fact that the payments of that small sum have been amounting in the aggregate to from £150 to £160 a month. About 10s. a week is allowed to each labourer on strike, and the mode in which a strike on any particular farm is got up is short and simple.

The men make the demand at the end of one week for an increase of wages the next. If, at the end of the next week, the farmer does not comply, he gets a formal notice that on the following Saturday the men will quit his employ should he still remain obdurate.

Some three weeks ago the body of men working on the farm of Mr. Hambridge, of Ascot, went out after due notice, and after he had declined to pay them the increase they demanded. There was a carter in his employment, who seems to have had no grievance of his own; but, as Mr. Hambridge had allowed his fellow-labourers to go out, the carter thought fit to join in the strike without any notice whatever. For this Mr. Hambridge summoned and obtained costs against him, and "the neighbourhood" added this to the previous cause of "aggravation" against the farmer. The next step in the business was the employment by Mr. Hambridge of two labourers named John Hodgkins and John Millan, whom he had brought from a village at some short distance from his own. The labourers out on strike took no visible part in any proceedings to obstruct the two strange hands; but the women of Ascot resolved on a physical force demonstration in favour of union principles. Some sixteen or seventeen of them assembled at a gate by which Hodgkins and Millan had to enter Mr. Hambridge's field, some of them being provided with sticks; and, saying that they were determined the men should not work for Mr. Hambridge, they "dared" them to enter on his land. Not a blow was struck, and though very abusive and threatening language was used, the fright and peril which a few girls and middle-aged women had occasioned to the two stalwart labourers could not have been very serious, seeing that, according to the evidence of the men themselves, the women actually offered to escort them back to the village and give them "a drink," and that, while declining the protection and hospitality of the Amazons who had so terrified them, they walked to Mr. Hambridge's homestead unguarded and unhurt, and subsequently went to work on the farm under the powerful protection of one police-constable. Most people will, perhaps, be of opinion that the matter might well have been allowed to end there; but Mr. Hambridge thought otherwise, and summonses were aken out at his instance against seventeen women. They were summoned under the Act 34 and 35 Victoria, cap. 32, which is "an Act to amend the Criminal Law relating to Violence, Threats, and Molestation." The first section of that Act makes it an offence within the scope of the enactment to "threaten or intimidate any person in such manner as would justify a justice of the peace, on complaint made to him, to bind over the person threatening or intimidating, to keep the peace." The case came before the Chipping Norton Bench on Wednesday, the presiding magistrates being the Rev. Thomas Harris and the Rev. W. E. D. Carter. Mr. Wilkins, a local attorney, appeared for the complainant. The women pleaded "not guilty," but they had no professional assistance. The charge as against one of them was dismissed; but of the seventeen charged, sixteen were found guilty, and the two reverend magistrates, after a "lengthy consultation," ordered seven of the women to be imprisoned for ten days, and the remaining nine to be imprisoned for seven days, with the addition of hard labour in every case.

Such a sentence staggered the poor women; and well it might, for it has staggered the whole county. The indignation at its severity is deep and out-spoken. It is fair to the two magistrates to say that the Act did not allow them the option of imposing a fine. If a fine had been inflicted, an officer of the Union was in attendance with money to pay it. The same section of the Act which enumerates the offences coming under it, provides that the offender " shall be liable to imprisonment, with or without hard labour, for a term not exceeding three months." But the justices had another option, which the law and usage gives all judges in such cases, and of which most people would have supposed any magistrates of ordinary prudence and humanity would avail themselves in this case. They might have allowed the women to stand out on their own recognisances, binding them " to come up for judgment when called upon." If the object of the justices was to prevent a recurrence of threatening, can any one doubt that it would have been effectually achieved by such a course? No one about the district appears to doubt it ; and the two justices have the less excuse for not having come to such a merciful and conciliatory decision because, if they had only been at the pains to inquire, they would have found that when, on the 9th of August last, some men were summoned before the Woodstock Bench under the very same Act, the justices then presiding adopted that course, even though actual violence, to some slight extent, was proved in evidence ; and that the consequence has been, that since then there has been no threatening or intimidation in the neighbourhood of Woodstock, and it has not been found necessary to require the men to come up for judgment. Mr. Carter, during the hearing of the case on Wednesday last, more than once asked Mr. Hambridge whether he really meant to press the case, and expressed the difficulty he felt in dealing with a few women under the provisions of so severe an Act. Perhaps the Rev. Messrs. Harris and Carter were not aware of the fact that they could let out the women on their own recognisances ; but it might have occurred to them that the women doubtless had committed the offence in entire ignorance of the Act 34 and 35 Victoria, cap. 32. At all events, with the Act before them, the two justices must have known that they might have left out the hard labour.

The labouring population of the village were astounded when they heard the sentence, but they bore it quietly until about nine at night, when the "roughs" of the neighbourhood, in which there is a manufactory, assembled in considerable force. Then, after much shouting, an onslaught was made on the police station; the windows and the door were broken, and some of the tiles were stripped off the roof. Police Superintendent Larkin and his men are admitted to have acted with great forbearance, but the superintendent thought it advisable to telegraph for assistance to Oxford, nineteen miles distant from Chipping Norton. On receipt of his telegram, Inspector Yates, with a force of police, started in a drag and four, and at Woodstock took up Superintendent Bowen. So riotous was the aspect of Chipping Norton, that it was not deemed safe to keep the women there till the time at which the first train leaves; and in the small hours they were driven in the brake the whole distance to Oxford, where, at about six o'clock, they were locked up in the county gaol. Two of the unfortunate prisoners had infants at the breast while being conveyed on their cold journey to prison. Petitions to the Home Secretary are spoken of in the district, and so threatening is the attitude of the village, that on Saturday evening police were again despatched from Oxford. The more respectable portion of the population believe that Mr. Bruce will feel it his duty to send down an order for the immediate discharge of the whole of the women, who have now been in confinement since Wednesday afternoon.

An intense feeling on the subject was stirred up throughout the whole country. The press unanimously condemned the decision of the Bench. Unhappily, the dilatoriness of the Home Office, to which petitions for the immediate release of the women were sent, prevented any governmental interference until it was too late. The Leamington committee of the Union took prompt action, and a member of the consulting committee, the Rev. F. Attenborough, in conjunction with Mr. Arch, at once issued an appeal to the public. The result was subscriptions amounting to eighty pounds, which it was resolved to give to the women on their release from Oxford gaol.

An interesting account of the release of these women was given in the current number of the "Chronicle." The following is an abridgment of it, the writer being a member of the consulting committee :—

As the prisoners were brought from Chipping Norton to Oxford in a conveyance, it was decided to take them back in the same fashion, and to give éclat to the proceedings, a handsome drag, drawn by four half-bred horses, was chartered from Oxford for the purpose. In this they left Botley shortly before twelve o'clock, amidst loud cheers from an enthusiastic crowd, which had gathered together in front of the Union office. At Woodstock the party baited at the King's Arms, and were well treated by the inhabitants. In fact, it was almost impossible for the women to move about on account of the shaking of hands and congratulations of the burgesses. Before leaving, Mr. Holloway conducted the party to Blenheim Park, and past the seat of the Duke of Marlborough, whose antagonism to the Union is most inveterate. On leaving Woodstock, there was loud cheering for "the Union," for President Arch, and the women.

Our ride from Oxford to Chipping Norton lay through

one of the fairest districts in mid-England. On either side the rich and fertile landscape unrolled itself like a panorama of matchless beauty. One thought, and one thought alone, marred the enchantment of the scene. It was the knowledge that behind all this luxuriance of natural wealth, there was wholesale human misery, wretchedness, and suffering. The loveliness of the scene only brought out in bold outline before my mind the terrible sketches of cottage life, the appalling ignorance of the rural mind, and the abject domestic wretchedness everywhere attendant on the agricultural labourer's existence. It was shown before the Royal Commission in 1869, that at Kidlington, the father and mother, two grown-up sons, two grown-up daughters, and three little children, all slept in one bedroom. The following narrative by Mrs. Bowerman, on the same occasion, is typical of the agricultural labourer's existence in this highly-favoured district:—"My husband is a milkman, and gets 12s. We pay 1s. 9d. rent, and have only one bedroom. Last week I got 3d. for a bit of sewing, so we had 12s. 3d., and I spent it in this way—meat, 3s. 7½d.; bread, 3s.; rent, 1s. 9d.; tea, sugar, flour, soap, and soda, 3s. 7½d.; boys' schooling, 2d.; total, 12s. 2d. So I had a penny left." Facts like these cast a dark shadow over the broad acres of Blenheim and Woodstock.

At Chipping Norton Junction an aged labourer recognised Mr. Arch, and in the course of the conversation related an incident in his own experience which will hardly be credited. It was to the effect that he was driven out of his native parish, by the squire and the parson, simply because he married without the consent of the former gentleman. The statement was made in such a manner as left no room for doubt, and from inquiries subsequently made, I feel satisfied that oppression of this character is still practised in remote country parishes.

COMMITTAL OF WOMEN TO PRISON. 141

The train which carried us from Chipping Norton Junction to Chipping Norton also bore there the nine women who were released from Oxford gaol on Tuesday last, whom Mr. Leggett had brought down from Ascot. They were on their way to join the other seven martyrs who had gone on in the drag. They were all neatly dressed, and conducted themselves with a decorum highly becoming their station in life. In the same train there was also a *posse* of police from Oxford, sent on to preserve order.

At Chipping Norton we found numerous bodies of agricultural labourers, who had travelled long distances to take part in the indignation meeting in the evening. Around the waggon I think there could not have been less than 2500 persons. Some of our friends said 3000, but I like to be within the mark. A better behaved and more orderly concourse of people I have never seen gathered together. The chairman, in opening the proceedings, advised the labourers to conduct themselves in a legal and orderly manner, and to see that there was no obstruction to any vehicles requiring to pass. He characterised the conviction of the women as unnecessarily severe, and said it was another instance of the necessity for reform in our arrangements for the administration of justice. The only satisfactory remedy would be (1) the extension of the franchise, (2) the repeal of the Criminal Law Amendment Act, and (3) the appointment of stipendiary magistrates. Mr. Banbury, sen., of Woodstock, moved, "That the Criminal Law Amendment Act is in its operation unjust and oppressive towards the labouring classes of this country : this meeting therefore calls for its immediate and entire repeal." In the course of a very earnest address, which was heard with evident interest, he reviewed the general bearing of recent enactments towards the working classes, and showed how unsatisfactory decisions tended to shake the confidence of the people of this country in the administration of justice.

With regard to clerical justices, he maintained that the Bench was not the place for a clergyman. Even if his decisions could be justified, they were sure to produce dissatisfaction in some quarters, and weaken his moral influence.—Mr. H. Taylor seconded the proposal. He gave several instances of obviously unfair decisions towards working men by the clergy.— Mr. Joseph Arch delivered a very powerful address in favour of unionism, the rights of the working classes to combine, the injustice of the Chipping Norton convictions, and the necessity there was for removing clergy from the Bench, both for their own good and also that of the public. The resolution was carried with acclamation.—Mr. Mottershead, of London, proposed, " That clerical magistracy is most unsatisfactory ; and, in order to secure a better administration of the law, it is considered absolutely necessary to establish a stipendiary magistracy, on account of class influence on the present administrators." —After Mr. Savage, of Birmingham, had seconded the motion, it was put and carried unanimously.—A collection was made in the crowd for the women, which resulted in upwards of £3. The sums given were principally pennies. Mr. Arch said he had received about £30, and hoped to be able to send £5 each for the women.

At a subsequent meeting this hope was realised, and, amid boundless enthusiasm, Mr. Arch presented each of the sixteen women with five pounds each.

A kind of counter-demonstration was got up by the sympathisers in the prosecution, which elicited the following trenchant rebuke from the labourers' organ :—

THE CHIPPING NORTON MAGISTRATES.

The "Times" and other papers have published the addresses of sympathy presented to the two clerical magistrates of Chipping Norton by some of their parishioners, and we read in the "Oxford Chronicle" their replies thereto. These delightful "parishioners" express "their entire approval" of the magisterial decision which consigned the sixteen respectable women (their fellow-parishioners, forsooth !—though there does not appear to be much advantage in being a fellow-parishioner) to the common felon's gaol. The clerical justices, in reply, actually accept these expressions of sympathy, approval, and support, as though they had fairly earned them by this notorious magisterial decision of theirs, which has excited so much indignation amongst their fellow-countrymen (we are glad there is at least some advantage in being a fellow-countryman); and possibly may thereby be encouraged under like circumstances to 'do the same again ! Oh, excellent, high-minded justices ! Oh, affectionate and devoted parishioners ! It is to be hoped that there are more enlightened parishes in some parts of England, where praise and blame are "laid on," like after-dinner speeches, as occasion may seem to require on the part of faithful parishioners, with such wonderful discrimination of the merits and demerits for which it is awarded.

No one doubts, we should think, that these gentlemen "endeavoured" to act to the best of "their knowledge and judgment;" but if it is shown, as we think has plainly been done, that in their magisterial capacity they have acted with very partial knowledge and most unsound judgment, they may be as totally unfit for such an office as if they had not "endeavoured" to administer justice at all. The consequences are just the same to the victims of "justices' justice," whether the convicting magistrates are right down bad men, or on the whole good men, but with imperfectly developed social sympathies. It will do these gentlemen good, if they are really good men, to undergo a little self-humiliation. If they are not good enough for that, they will flatter themselves they have been treated unjustly, and won't profit by the lesson they have had. In the latter case, their removal from the bench is an imperative necessity. In any case their removal would be a good thing, as it would teach "the great unpaid" the most salutary lesson that they are responsible to a broad and enlightened public opinion, and not to their own personal friends and parishioners, for administering justice with an equal hand and without class prejudice. Fancy the two Chipping Norton clerical justices committing sixteen farmers' wives for so slight an offence as that for which they unhesitatingly sent sixteen labourers' wives to prison!

Can any of the parishioners of the Revds. Carter and Harris possibly conceive such a thing? And supposing all the labourers of the parish had presented addresses to the rev. magistrates, expressing approval of their conduct, notwithstanding that the almost unanimous voice of their countrymen generally had condemned it, should we not think that magistrates who could quietly accept such expressions of approval, as though they had been grossly ill-used men, were utterly incapable of profiting by the rebuke they had received, and were therefore confessedly unworthy of being entrusted with such responsible duties again? The least that could be expected of the Chipping Norton magistrates is that they would confess their error, as professedly Christian men, and in turn rebuke their parishioners for supporting them in what they ought by this time to have understood to be an instance of cruel injustice and class prejudice and oppression. The clergy might as well shut up their churches and doff their sacred robes, as not be capable of genuine sympathy with the poor. They have certainly no place on the magisterial bench, however they may be tolerated as representatives of a State Church.

The next event of interest in connection with this movement, was the second annual conference at Leamington, which took place on the 21st May, 1873. Mr. George Dixon, M.P. for Birmingham, presided, and a large number of delegates from the rural districts were present. A considerable number of the outside sympathisers in the movement also attended. The report stated that the National Union consisted of twenty-three district unions, with nine hundred and eighty-two branches, and seventy-one thousand, eight hundred and thirty-five members. Twenty-four counties were embraced in the movement.

The funds of the Union were reported in a somewhat exhausted state, owing to the vigorous but unsuccessful attempts on the part of large numbers of employers to stamp out the Union by means of "lock-outs" among their unionist employés.

A considerable amount of liveliness was infused into the meeting by a personal altercation between Mr. W. G. Ward and Mr. Edward Jenkins, two members of the consulting committee and trustees of the Union funds. The difficulty arose out of a strongly worded article from the pen of the former gentleman, which had appeared in a recent number of the "Chronicle." As this controversy has occasioned considerable excitement among the patrons of the Union, I will give the article of Mr. Ward *in extenso.* It appeared in the number for May 10th, 1873, and was as follows:—

THE BATTLE, THE STRUGGLE, AND THE VICTORY.

The struggle of classes has commenced; the strife is deepening daily, and the battle is very near. The farmers have determined that the farm labourers shall not combine, that their serfs shall continue serfs, their claims to manhood crushed; that their wages shall not be a matter of mutual consideration' or market value — the labourer shall have a maximum of wages at the dictation of a combination of farmers. He shall not be consulted, or his feelings and demands considered: submit or starve. And this tyranny in England in the year 1873!

Yes, to-day, while the tenant farmers are agitating for tenant rights, they are trampling upon the rights of their labourers; while they are combined to get legislative protection, and by legislation, public money to assist them in their business, they are bitterly indignant that their serfs should presume to combine, even to attain free agency to the simplest element of manhood—the right to say what their labour is worth.

In Essex and Suffolk and elsewhere there is a Tenant Farmers' League to destroy the combination of the labourers, binding themselves to each other to refuse employment to all labourers in union, and not to give more than 2s. for a day of twelve hours' work; and now the farmers have locked out, in various parts of the country, fifteen hundred men, and cast them out to starve, or what they expect and aim at, to eat up the Union funds and leave combination stranded. The farmers foolishly imagine that with six weeks' comparative leisure before them, they can fling their men upon the Union funds and exhaust them, destroy combination, give their men a sharp pinch, and get them back broken in spirit, humbled to submissive serfs, to work for them

THE BATTLE AND THE VICTORY.

just as they want them, to mow and moil for less than a slave's pittance, with no outlook of hope, no escape but in the grave.

Have the farmers no power of reflection, no memory of agrarian anarchy, no conception what manhood is, when tyranny is striving to crush out manliness, and tramples upon simple but hearty hopes, and leaves no alternative but slavish submission or manly rebellion? Cannot the farmers calculate the effects of their success in the exhaustion of Union funds, the prostration of the peaceable agitation, and the dispersion of the Union delegates and their leaders? Do they know so little of their men, know nothing of the character of Englishmen, that they believe that, silence the Union, and the men will tamely submit to the yoke of slavery; and that beat back Englishmen from the glimpse of freedom, tear up their union cards, and unity is gone, starve them a week, and their submission is perfect? Bah! Lock up the foolish farmers and let the mad go free.

If our labourers inherit an atom of the spirit of their forefathers, of those who fought at Cressy and Waterloo, and struck for freedom at Naseby and Dunbar, they will not basely bend the neck to a farmer's yoke, let the consequences be what they may; if our labourers inherit within their breasts the hearts of iron of their forefathers, that gave them courage to face the battle with any odds, to bravely force the breach against any numbers, to win for their country in a hundred battles, surely now they will not sneak away at freedom's call, and rush to dishonoured graves, frightened at the frown upon a farmer's brow! Shall they, whose forefathers in the past gave their blood and lives freely for their country, not be able now to give all to win one battle for themselves?

The farmers have blundered; they have thrown down the gage of battle to their men, and come what will they will be beaten. The farmers have done worse than blundered—they have insulted every Englishman and sought to tarnish the fair name of our country; they have treated their labourers as aliens in blood, as unworthy the dignity of freemen, or even the nobleness of manhood. Though our farm labourers are, outside the constitution, treated simply as beasts of burden, yet they have a legal right to combine for mutual protection. But no, say the farmers of Essex and Suffolk and elsewhere; if we have power you shall starve if you combine; you shall not dare to stand upright and bargain with us for wages; as the slave accepts submissively what his master condescends to give him, and the dog his bone, so shall you our 2s. a day, and the day shall be long if the money is short.

The farmers have ever blundered. They have always gone upon the penny-wise principle, centred in self, seeing only one side, and that their side, blind to the future, incapable of embracing their country, much less humanity; incapable of considering secondary consequences, and therefore often astonished at the evils of their own success, and to find that a trumpery victory involves irretrievable loss; and as no man can be a tyrant who is not at heart a slave, the farmers have matriculated under the gamekeeper, the landlord, and the parson, to learn unmanly submission, to qualify themselves to be tyrants over their labourers, and try to crush the man into a serf.

If the farmers were now to succeed by their lock-out in exhausting the Union funds and prostrating the peaceful agitation, then they would have to face far greater difficulties. They would find they had suppressed a moderate and legitimate agitation, only to arouse agrarian anarchy; they would find their docile labourers, deprived of their Union leaders and their combination crushed, completely changed. Maddened by tyranny, infuriated by lawful progress, barred, driven to desperation, they would feast upon revenge and glory in anarchy. And who can be surprised? Blood is thicker than water. And an Englishman that would tamely submit to class tyranny, to a combination, unjust and unmanly, to crush a legal and noble combination of a depressed and down-trodden class, seeking true manliness and social elevation, would deserve to be spat upon as a willing slave, to be scouted as an outcast, and dungeoned in darkness, until the fire of freedom burned in his breast, and he could offer himself a sacrifice on the altar of freedom.

The farmers may depend upon it, that if they could succeed in crushing the combination of their serfs, in scattering the Labourers' Union, the success would be signalled by midnight surprises, by beacon fires, from one end of England to the other. There are circumstances that justify war, even civil war. When the oppression of five hundred years had apparently degraded the peasantry of England into brutish insensibility to manliness and freedom, then thoughtful men sighed and lamented at the hopelessness of their elevation, from their stolid indifference to their own degradation. But all in a moment the indifference and insensibility were gone; the serfs awoke and claimed their dues with calmness and moderation, with a business shrewdness, and a wide comprehension of circumstances, and a statesmanlike wisdom, and often with simple but dignified eloquence, that charmed as much as it surprised all who believe in progress, and desire the moral elevation of their countrymen. Then the leader of the labourers was interviewed by statesmen, helped by philanthropists, feasted by mayors and members of parliament, and patronised and smiled upon by thousands of Englishmen, from the artisan to the manufacturer, who were gratified to see that the English serfs, hitherto the shame of their country, manifesting that with fair play they could redeem themselves and shed honour upon their country. Twelve months of Union education has given to the serfs the notions of manhood and freedom, the knowledge that union is strength, and that sober and industrious men deserve the franchise ; and their eyes are opened to the evils of the land monopoly, of vicarious cultivation, and the necessity of a wide diffusion of the land amongst the people, and now they are fully awake to combination, to the freshness of the air of freedom. A handful of farmers have banded themselves together to arrest the flood of freedom—to stamp out combination by counter-combination—to overcome evil by evil, and rivet the chains of serfdom upon their labourers, by fixing a maximum of wages, insisting upon their combination being broken up, that starvation and lock-outs shall involve submission, and their hopes shall be buried at the dictation of their masters. Surely such tyranny would justify resistance, even to blood and war.

Men of England, artisans of our towns, manufacturers, traders, shall it be that the farmers shall drive back to serfdom the growers of your food ?—shall they succeed in again trampling down to semi-starvation our poor farm labourers ? —shall it be that in the richest nation of the world, the peasantry shall be the meanest, most debased, and most down-trodden of all the cultivators of the soil, in all the civilised states of Europe and America ?

We appeal to you, as help is required. The funds of the Union are going out £500 a week over the receipts to maintain these poor locked-out labourers. Migrate them across England, and help them to emigrate, and seek in a foreign land the bread that is denied them in their own. We know we shall not appeal in vain. We know the heart of England is with us, and no Englishman, except a farmer or a landlord, will stand supinely by and see the farm labourers driven by desperation to wild revenge and furious anarchy.

But if there is no alternative between submission to serfdom and starvation, if civilisation has no cheer, no comfort, no protection for the farm labourer, can he be expected to tamely lie down and die? No, no. Better to die in a noble struggle for freedom than linger out a degraded life, a farmer's chattel, than creep on slowly, hunger-bitten and tortured, to an ignominious pauper's grave; better far curse a one-sided civilisation that worships wealth and crushes honest poverty and die, than live and struggle without hope, toil till despair paralyses all power, and leaves a withered heart and racking frame to the tender care of a board of guardians, composed of farmers, to doctor, damn, and bury.

Men of England, artisans, miners, merchants, traders, we appeal to you. Shall tyranny triumph, shall combined farmers lock out and starve our peasantry into submission?

If we had any doubt of you, despair and desperation would redden our villages, and anarchy would triumph over selfish luxury and brutal tyranny.

Farm labourers who are in work, stand by your Union in the dark hour of its struggle for your freedom, earnestly and faithfully as a mother watches the sick bed of her only child. Every man who owes a single copper, pay it up; volunteer a levy if you will; every nerve of your leaders will be strained for you. Money must be found, and our towns will be visited, and meetings called, and a hundred town-halls shall ring with cheers for the down-trodden farm labourers. On the moors of Durham the colliers shall meet and hear our appeal; and on the banks of the Tees shall the iron-workers back us with a hearty response.

Farm labourers, never fear, stand with your face to the foe, and he shall yet quiver, smart, and flee:—

> "For freedom's battle once begun,
> Bequeath'd by bleeding sire to son,
> Though baffled oft, is ever won."

The night may be dark, but the morning star of hope is visible; and soon, very soon, the sun in the radiant glory of light and warmth of perfect freedom shall shine around you.

Farm labourers who are locked out, cheer up; suffer anything rather than give way to tyranny and oppression. Obey the Union delegates—attend to committee orders; be patient and faithful, and you need fear not. Every penny of the Union funds necessary will be spent to sustain you in the struggle, and before two months are past your tyrants shall be humbled.

But remember you have to bear a share of the burden. The Union may have to send you a long way, and the accommodation may be scanty, but the wages will be good. Do not pine and moan if the bed is hard, and provisions have to be fetched some distance. Bear anything, bear everything, rather than bend your necks at a farmer's selfish, unjust dictation,—rather wear out in struggling that you may leave your children a heritage of freedom, and not that you and they shall for ever be serfs, have your wages fixed without consulting you, and be compelled to sign one-sided agreements to keep you in bondage for a year, and leaving your taskmaster free to dismiss you without any penalty, but for you a jail if you dismiss him; and never bend to masters who combine against you but will not allow you to combine for mutual protection.

England expects every man to do his duty. The trustees, executive and consulting committees, delegates and secretaries, are at their posts. The towns will be appealed to as soon as necessary. The organisation is complete, AND VICTORY SHALL BE OURS. He who would succeed must determine success shall be his.

Men of England, be with us in our struggle, that the victory may be with the oppressed, that serfdom may cease in England, and her bread-growers receive the fair reward of their labour ; that hunger and oppression may no longer dog the steps of the poor farm labourer.

THE LABOURERS' UNION FOR EVER ! !

A few days after the appearance of this article, the following letter appeared in the "Times" :—

AGRARIAN AGITATION.

To the Editor of the " Times."

SIR,—The following passage appeared in the "Labourers' Union Chronicle," published at Leamington, on the 10th inst. :—

"If the farmers were now to succeed by their lock-out in exhausting the Union funds and prostrating the peaceful agitation, then they would have to face far greater difficulties. They would find they had suppressed a moderate and legitimate agitation, only to arouse agrarian anarchy ; they would find their docile labourers, deprived of their Union leaders and their combination crushed, completely changed. Maddened by tyranny, infuriated by lawful progress barred, driven to desperation, they would feast upon revenge and glory in anarchy. And who can be surprised? Blood is thicker than water ; and an Englishman who would tamely submit to class tyranny, to a combination, unjust and unmanly, to crush a legal and noble combination of a depressed and down-trodden class, seeking true manliness and social elevation, would deserve to be spat upon as a willing slave, to be scouted as an outcast, and dungeoned in darkness, until the fire of freedom burnt in his breast, and he could offer himself a sacrifice on the altar of freedom. The farmers may depend upon it that if they could succeed in crushing the combination of their serfs, in scattering the Labourers' Union, the success would be signalled by midnight surprises, by beacon fires, from one end of England to the other. There are circumstances that justify war, even civil war."

This passage has naturally given rise to severe comment in the provincial press; it is entirely opposed to the aim and spirit of the labourers' movement, as I have seen it manifested by Mr. Arch and his coadjutors from the beginning until now: it is, moreover, compromising to the large Consulting Committee, containing the names of Mr. Dixon, M.P., Mr. Morley, M.P., Canon Girdlestone, and others.

Will you permit me to explain that the "Labourers' Union Chronicle" is a private venture, and has no official connection with the Union; that it is not under the control of the Executive Committee, nor are they responsible for anything that appears in its columns, unless it is signed by their officials or put in as an advertisement. And, moreover, I am happy to say that this indefensible piece of truculence does not come from the pen of any labourer or artisan, but of one who is a landowner, a country gentleman, and a person of cultivation.

I have never heard such words from Mr. Arch or any country labourer, and regret that a good cause should be injured by advocacy so zealous, but detestable; and were I not sure it would be repudiated by the men themselves, I, for one, would withdraw from the Union.

I am, sir, your obedient servant,

May 20.   EDWARD JENKINS.

This letter Mr. Ward resented as a truculent impertinence; and, amid considerable impatience on the part of the audience, proceeded to state his grievance. Through the intervention of mutual friends, the affair was ultimately amicably disposed of.

A large public meeting was held in the Circus at Leamington, in the evening, Mr. George Dixon presiding.

The interest and importance of this meeting will, I think, justify the copious report of it which I have embodied in my "sketch."

The principal speeches were by the chairman, Mr. Allington, a labourer, Mr. Edward Jenkins, the author of "Ginx's Baby," and now M.P. for Dundee, Mr. Jesse Collings, a Birmingham town councillor, Mr. J. A. Campbell, a justice of the peace, and Mr. Joseph Arch. They were as follows — taken, I ought perhaps to say, from the shorthand report in the "Labourers' Chronicle" for June 14th, 1873 :—

THE PUBLIC MEETING IN THE CIRCUS.

At the conclusion of the business of the congress on Wednesday, a public meeting was held in the Circus, under the auspices of the Union. Mr. George Dixon, M.P. for Birmingham, presided, and with him on the platform were the Rev. Arthur O'Neil, Dr. Langford, Rev. F. S. Attenborough, Messrs. J. A. Campbell, Jesse Collings, J. C. Cox, E. Jenkins, Joseph Arch, H. Taylor, Haynes, Allington, W. G. Ward, and a number of other friends and supporters of the movement. The building, which is estimated to hold 2,000 persons, was well filled, and amongst the audience were many of the tradespeople of the town.

The Chairman having briefly explained that the commencement of the proceedings at this meeting had been unexpectedly delayed by the business of the conference, went on to state that he had been asked to open the meeting with a few remarks bearing upon the subject for the discussion of which they had assembled together. He was very glad indeed to be with them once more, after what he called a year of very successful agitation on the part of the Agricultural Labourers' Union. A short time ago a noble lord residing in a neighbouring county made some remarks upon their movement, and in the course of his address warned the agricultural labourers against the very serious undertakings in which they were involving themselves. His argument was this :—That the semi-feudal and semi-parental relations which had existed between the employers and the employed in the agricultural districts, were now going to be seriously disturbed, and perhaps it might not prove to be altogether to the advantage of the agricultural labourers to substitute in their place those ordinary commercial relations which obtain in large towns. Now he (Mr. Dixon) thought those remarks were very just, and very well worthy the careful consideration of the agricultural labourers. It was well, before a great movement of this kind was commenced, to count the cost, and to learn what it was they were going to do, and what they might have to suffer. He believed the meaning of the argument he had just given to be this :—If the agricultural labourers were to unite, and to say to their masters that they thought they were entitled to certain money wages and certain other conditions, such as an abridgment of the hours of labour, the reply on the part of the employers would be that the privileges hitherto offered to the labourers, partly in consequence of their low wages, would be partially, and in some cases, entirely withdrawn. Well, that no doubt would be to a great extent true. What those privileges were, and what was the proper value to be set upon them, the agricultural labourer was best able to determine ; but it was nevertheless a curious fact that exactly in proportion as these commercial relations did exist between the agricultural labourer and his employer, or in other words, exactly in proportion as the agricultural labourer found himself placed near to the large manufacturing towns and mining districts, exactly in that proportion was his position improved. Now, he had had an opportunity of ascertaining, on the very best authority—on the authority of a clergyman of the Church of England, now resident in the east part of Somersetshire—what these semi-feudal and semi-parental relations were, and what was the character and extent of that care which the farmers and others exercised over their labourers. The gentleman of whom he had spoken was very well able to estimate the advantage to be derived from them, because he had previously lived the greater part of his life in a manufacturing district, and therefore possessed some knowledge with regard to both sides of the question. He said it was a very curious thing, that although in that part of the country the great product of the land was milk, which was extensively used for, making cheese and butter, and although it was obvious that in the rearing of a family nothing was more valuable than milk, yet milk was the most difficult thing for a mother to get for her children. It had to be begged as a great favour, and was not always granted. He thought that semi-parental care of that kind did not amount to much. That clergyman further told him he was of opinion that the allotment system was a good one, and by way of giving a practical expression to his views, he determined on setting apart a portion of his glebe land for the purpose of allotting it out to the labourers at the same rate of rent as was paid by the farmers. The moment, however, the farmers heard such a thing was intended, they were loud in their protests against the system. They warned him that he would not get his rent paid, and went so far as to say that if the labourer gave attention to his garden, he would have less strength to devote to the duties of his master, and consequently would not be so good a servant. The clergyman, however, persevered in his object, and he had since stated that the labourers, one and all, considered these allotments a great boon, and none of them had ever fallen in arrears with their rent, no matter what the state of the crops had been. That gentleman also thought it was essential for the well-being of the agricultural labourer that he should be educated, and in accordance with that view did everything in his power to induce the farmers of that district to assist him in the building of a school. But he was unable to induce them to do so until the Elementary Education Act was passed, since which time they had been compelled to subscribe. In neither of the cases cited did it appear to him that there was much benefit to be derived from that semi-feudal and semi-parental relationship existing between the farmers and the labourers, and which was said to be endangered by the present agitation for higher wages. They all knew very well indeed that the wages in that part of the country (East Somersetshire) were very low indeed, perhaps amongst the lowest that were paid to the labourers in any part. Now, what was the result of that? The clergyman of whom he had been speaking explained to him that two men out of Yorkshire would do as much work as three men in East Somersetshire, so that if a farmer employed two of the former at 15s. a week, he would get more work done than by employing three of the latter at 12s. a week. What did that state of things mean, save this, that the small wages given the Somersetshire men was without doubt the cause of the insufficiency of their labour? If the farmers in that part of the country would do what the men asked them to do, namely, discontinue the old relationship and begin a new one, in which a fair day's work should be rewarded by a fair day's wage, they would find their labourers lifted up to an equality with their brethren in the north, and well worthy the increase of money paid them. Under the old system it was not considered the agricultural labourer was a fit person to have the franchise, consequently, the franchise had not been given him ; and so long as he came to be regarded as the fitting object of the kindness and charity of those who were above him, so long those above him would demand that in return for that kindness and for that charity he should continue to be their political dependant. His (Mr. Dixon's) opinion was that under the new system, which had been fairly described as a commercial system, the agricultural labourer would gain that degree of independence which would enable him to forego the advantages which had hitherto been derived from the "semi-feudal and semi-parental" relationship heretofore existing between employers and employed in the rural districts. At the same time he did not wish to hide, neither from them nor from himself, the fact that in the change, which was one of enormous magnitude, they would have to undergo many trials, would have their patience tested, and their powers of endurance tried. But what had they done already after one year of trial? They all knew how largely wages had been increased, in some districts 1s. per week, in others 2s. per week, in others 3s. per week, and he had heard of one district where the increase had been as great as 4s. per week. In a speech lately made in the south by Mr. Arch, he quoted the opinion of a gentleman in whom he placed great confidence, to the effect that the aggregate increase in wages during the year amounted to no less than £1,000,000. Now, he did not wish to endorse that statement, because he had not the materials necessary for forming an opinion, but they knew that the sum was very large, and that it was continually increasing. That, however, was not the only good that had been derived from their labours during the past year. Thousands of those who had been under the most depressing circumstances, unable to rise, and with no hope before them, had been removed into districts in England where wages were higher, where openings of various kinds surrounded the working man, and where he believed all had bettered their position, and some would eventually rise to be men of wealth and substance. Besides that, very great attention had been paid to the question of emigration, and many had been assisted to emigrate. There had been a great number of persons prejudiced against emigration, partly because of a dread to go across the water, while amongst others a well-grounded prejudice had arisen out of the reflection that if our laws and customs were what they ought to be, and it was hoped they would one day become, there was ample employment on remunerative terms for all our labourers at home. Whatever might be the soundness of that opinion, it was clear that by encouraging the labourers to leave the country now, they were hastening the time when those labourers would be wanted, and not being then at hand the country must suffer. He was one of those who believed in the opinion that if we had wise laws, and if the spirit of those laws was carried out into everyday life, such a result would be impossible. But his experience as a politician was that they would have to wait a long time for these wise laws, and they would have to wait a still longer time until the spirit of the people had comprehended the wisdom of such laws, and carried them into execution. What was to be done in the meantime? As one who had been in the English colonies, he could tell them how fully he appreciated the advantages held out by Canada, and Australia, and New Zealand, as fine countries as England, in some respects much better, while in no respect were they really worse. There there was magnificent land, almost to be had for the asking. If any portion of those extensive countries could be possibly detached and tacked on to Land's End, so that emigrants had only to pass over a bridge to gain possession, thousands upon thousands would leave and go over at once, so that soon there would not be an acre to be had. He could not see that emigration would be a great evil to England, seeing that every one who went to the colonies,—every agricultural labourer, who went out in the spirit of industry and in the spirit of independence, went forth to certain wealth. In the "Times," the other day, he came upon a letter from some one in New Zealand, who said that one hundred labourers had arrived in Christchurch, and met with remunerative employment in the course of two or three hours; and there were five hundred families in that part of the country desirous of obtaining such labourers as were then present, and could not obtain them because they were not to be had. One family that had emigrated, in the course of six or twelve months had saved £90. It was obvious that every man who could muster up courage to cross the water, would materially benefit himself in either of the colonies to which reference had been made. And how about those who remained behind? The number of labourers had been said to be so large that wages were depressed on that account. Well, if this stream of emigration was kept up, there would of course be less competition at home, and eventually labour would become so scarce, that in case of a strike it would be absolutely essential for the farmers to give the increase of wages. Let them remember, that it very frequently happened, when persons began really to understand the nature of what was going on about them, they did not wait for accomplished facts. So if the farmers came to see that some of the men were migrating to the north of England, and did not come back, and that others were going across the waters and did not come back, they would soon begin to see that this was a very serious matter—that it could not be allowed to go on, and that some steps were necessary in order to make the men satisfied and induce them to stay at home. They would say to one another,

"Let us repair their cottages; let us give them a little land — even enough to keep a cow and a pig; let us try to better their condition, and even go so far as to increase their wages." Let it be remembered, too, that the emigrant removed to New Zealand still remained an excellent customer to this country, consuming much more of our commodities than he did when an agricultural labourer at home. Bearing in mind these facts, he looked hopefully into the future, because he saw all these things taking place, and however certain men may gloss them over, he said, when gazing upon these facts, that they were facts of extreme significance, and would eventually thrust themselves upon the notice of even half-blind men. And now he would ask them to look upon the other side of the picture. Forbearance was necessary towards the farmers. Very great consideration was required for them. Their position was one of peculiar and great difficulty. It had been said (but he did not know with what degree of truth, for it was one of those things that never could be ascertained), that they had had a bad year. He knew farmers always had bad years. He was told that it was very much more difficult for farmers to get good labourers than it used to be. Part of that difficulty no doubt arose from the movement now going on amongst the labourers. What the farmers felt was uncertainty as to the future; and he might say that the labourers did not intend stopping simply with an increase of wages. When they had obtained a settlement of the wages question, they would ask, "How about the hours of labour?" The farmers, therefore, were asking themselves how long this state of things was to go on, and also whether they could stand it, because they argued that, as the price of corn was regulated not by the amount of produce, but by the export trade, they were unable to increase their selling rates in proportion as the cost of production was augmented. Therefore it would be seen that the position of the farmer was one of great difficulty, and he confessed that it excited his sympathy for them as a class. If the farmer could go at once to the landlord and say, "My position is such that it is impossible for me to work my farm, unless I have the rent reduced," and the rent was then lowered, the solution of the difficulty would be easy. But it was a very difficult thing to prove to a landlord that the farm could not be worked without a profit, and that a reduction of the rental was an absolute necessity. Generally, the position of the farmer was by no means so satisfactory as some of them could wish. The fact was, there was now going on through the agricultural districts a change which was little short of a revolution. And what they had to say to the farmers and landlords respecting the matter was this, "We admit you are going to be placed in a position of difficulty, but although the sacrifice you may be called upon to make will be a great one, the good that is to result from it—the elevation of an entire class—is greater than the evil." He would also say this, "Put aside all those old-fashioned notions about looking after your labourers, and treating them as you do your cattle—look upon them as men, and place them in a position of independence, and take care that they shall be educated, so that they will be able to do your work better, with better machinery. Give them such wages that they will be physically developed to the extent of being able to do what they now do, or twice the work they now do. Educate yourselves — and then it may be that the new system of working the land may be so improved that you will find that you will be able to do with perhaps fewer workmen, with higher class workmen, pay greater wages, and be able to make more profit out of your land." He felt no doubt such would be the case, but in order to get the land laws altered the labourers must first have the vote. He looked forward to the time when many of them would become, landowners it may not be, but the holders of the soil, when they would be able to work the land for themselves. When the day came that the land of the country should be thrown open, and the transfer of it made easy and cheap, he had no doubt its value would rise to such an extent that the landlords would be greatly benefited by the change. These great changes were necessarily slow in their operation, and sometimes involved individuals in a certain amount of trouble, but in the end they contributed to the happiness and prosperity, not of one class alone, but of every class in the community.

Mr. Allington said he felt great pleasure in appearing before them on this the first anniversary of the National Agricultural Labourers' Union, and especially was that the case, as he was not present with them at the meeting twelve months ago. As soon as Mr. Arch raised his voice at Wellesbourne on behalf of the down-trodden sons of the soil, he (Mr. Allington) commenced his labour for the cause by organising a branch in his own village. When he first heard of the movement, the good-tidings made his heart leap within him, and opening up communications with Mr. Arch, he received directions to commence his work. Well, for his exertions in the cause, he lost his situation, and afterwards went out for the Union as a delegate. With friend Harris, he visited Dorset, and in nine weeks organised several branches, and enrolled upwards of 2,000 members. Since then he had laboured in nine or ten counties successfully, as many could testify that evening. He had also been engaged in seeking employment for labourers who had been discharged for joining the Union, or were desirous of improving their position. In this way, he believed, much good had been done. The labourers who had been thrown upon the Union funds had been quickly absorbed in the manufacturing districts, to their own advantage and that of their families.

The result of this migratory movement had been satisfactory. With feelings of extreme pleasure he bore testimony to the readiness of Trades Unions in all parts of the country to assist, as far as they were able, the efforts of the agricultural labourer to improve his position. The general feeling was one of determination that the agricultural labourer should not sink. Of course the main strength, and the secret of strength, rested with the agricultural labourers themselves. But so long as they remained true to the cause and to one another, and so long as they continued to strive for their rights, he believed there were thousands of trades unionists in all parts of England who would never let them stand still for want of funds. Recently he paid a visit to Nottingham, when a committee was formed amongst the working men to collect funds for the Union. The result of that, he was happy to inform them, was the realisation of upwards of £100. They also had some boxes made for the principal workshops, and on the previous Saturday night, the pence and silver dropped in by the workmen amounted to £33 17s. These facts showed how the sympathies of the artisans in towns were in harmony with the Union, and how ready they were to assist the cause liberally and promptly. In conclusion he moved the subjoined resolution, and at the same time observed that if the members of the Union continued working together harmoniously, he was sure their work would be successful :—" That this meeting rejoices at the success which has attended the National Agricultural Labourers' Union, and, whilst deprecating the hostile attitude of the farmers and landowners to a legal combination of labourers, hopes that the Union will continue with firmness the policy of patient and united determination to redeem the agricultural labourer from the degraded position he has so long occupied."

Mr. E. Jenkins seconded the motion. He said, whilst his

friend Mr. Allington had been addressing them, he (Mr. Jenkins) had been envying, not his powers of utterance, but his legs. He only wished he possessed such a pair of agricultural legs. For those legs he was willing to change places with Mr. Allington at that moment. He felt sure with all the strength he had, the Union may look forward and count upon his services for at least thirty years to come. They had heard from the chairman the argument respecting the "semi-feudal and semi-parental" relationship existing between the farmer and his labourer. The statement reminded him of a young Irishman, who had been living in the north of England, where he got into some swell society. In the course of his conversations he never made any reference to his parents, of whom he seemed to be somewhat ashamed. One of his friends saw his father on one occasion, and afterwards meeting with the son said, "I think I saw your father the other day?" The young Irishman replied, "Well, he is my father to a certain extent." Well, the parental relationship existing between the farmer and the labourer appeared to him to partake of that qualified character—it was only "parental" to a "certain extent." He believed if they appealed to the farmer, and claimed his friendship as a parent, he would turn round and say, "It's all very well, my friend and brother, but our relationship is one only to a certain extent." To "a certain extent" they ought to feel an amount of sympathy with the farmers. What was their position? It was that of being between two grinding stones. They were like the ham in the sandwich; they were in the middle, and when the sandwich was bitten the ham was sure to get it. He knew there were some (and he was not quite sure the chairman did not once belong to them), who, with reference to the land question, had some doubts as to whether it would not be better to allow the plethora of labour to go on increasing until we had arrived at such a state of things that we could endure it no longer, and then it would work out the question of land reform. But for himself he would say that he entered into the emigration movement years ago, with an amount of energy he did not now possess, because he believed it to be a good and noble work, and a step in the right direction. He had always regarded our colonies as the proper heritage of the working men of England. He and other gentlemen, as well as Mr. Arch, had all been accused of setting class against class by aiding this movement. He remembered being in the House of Commons when Mr. Laird, of Liverpool, was charged with the responsibility of having built the *Alabama*. To that accusation he replied that he would rather be the builder of ten *Alabamas* than be guilty of going about the country setting class against class, at the same time pointing to Mr. John Bright, the member for Birmingham. Since then the hon. gentleman had taken his seat in her Majesty's Government, and he (Mr. Jenkins) had no doubt there were some people in Birmingham who regarded him somewhat in the light of a Conservative. After an allusion to the mischievous effects arising from class legislation and class magistrates, Mr. Jenkins stated that in the early part of this movement the "Times" refused to insert an advertisement from the Union, on the ground that its object was to set class against class. Considering the amount of opposition which had been brought to bear upon this movement, the success it had achieved was something wonderful. From one end of England to the other, whatever justices might do, and whatever might be the tone of the farmers, the taunts of the clergy and the Tory press, he hoped every member of the Union would remember that the movement was now getting thoroughly organised and must succeed.

The resolution was supported by Mr. J. A. Campbell, J.P., who said, if Mr. Jenkins sincerely regretted that he had not the "understandings" of Mr. Allington, what did they think were his feelings on that subject, seeing he was a cripple! He was not, however, going to make a "lame" excuse on that account. Looking back at the past year, he confessed his astonishment at the success of the Union; it was so much greater than he had anticipated. During the day he had heard some very strong observations about justices, farmers, and landowners, and being himself all three, he could say he was not ashamed to be any one of the three. During the time he had been engaged in farming operations he had employed a considerable number of labourers, and from what he had seen, his opinion was decidedly in their favour; so much so, that as soon as the agitation broke out, he felt it to be an important question as to what course he, as a farmer and a justice, should take. Accordingly he attended a meeting at Dunchurch, whereat the question was discussed by Mr. Arch, and he at once decided that it was his duty to favour the movement, because he was satisfied the cause of the labourers was honest, just and upright. He would have been heartily ashamed of himself as a Christian man if he had not come forward and done all that lay in his humble power to assist what he believed to be a thoroughly proper movement in this country. There were, as they well knew, different sights amongst different people. There was, in the first place, the "short sight," and then what was called in his own country the "second sight." There was also the "long sight," by which persons were enabled to see objects at a long distance, as contradistinguished from the "short sight," which was the ability only to perceive things close at hand. His opinion of the farmers was, that their policy was a "short-sighted" one, inasmuch as they did not ask themselves whether there was not some foundation for what was now taking place in the country. Viewed from any standpoint, they would arrive at the fact that the state of matters with regard to the agricultural labourers of England was not what it ought to be, and could not therefore possibly last. Then what should be the conduct of those connected with the labouring classes? Surely common sense ought to tell them, that as this change was coming over the country their wisest course was to meet it. But they ought not to meet it with the old idea that they could repress it with force, but meet it in a spirit of kindness and forbearance, and give to the claims of the labourers that consideration to which they were so fairly entitled. What was now taking place in the country had been rightly characterised by the chairman as a "revolution," but it was important to bear in mind that it was a peaceful and not a bloody revolution. They must all be short sighted if they failed to see that the agricultural labourers were taking proper measures for elevating themselves to that position they ought to occupy. He had been asked to become a member of the Consultative Committee, and had readily acquiesced, because he considered the men were deserving of sympathy. Since meeting with them on various occasions, he had been surprised at the great amount of really good sense and sterling judgment, accompanied with extreme moderation, he had witnessed on their part when assisting in the business of the Executive Committee. The fact was this movement was well founded, and if people would throw themselves against it with violence, probably it would end in their own injury and destruction. All the agricultural labourer had to do was to move on quietly, steadily, and perseveringly, and he must inevitably find his proper place amongst the people of this country. What was that proper place? Simply that he should be a man amongst men. The agricultural labourer was as much a man as the greatest nobleman in the land. He had his feelings as other men, and knew when he was treated well and when he was treated ill. It would be odd if he did not. He was capable of choosing his labour, and possessed a perfect right to dispose of that labour in the best market at the best price. If the agricultural labourer were, however, to attempt to force up wages higher than the farmer could afford to give, he would be doing a very foolish thing. He had already mentioned to them that he was a justice of the peace, and having done so, he felt he could not sit down without saying a few words with reference to that most extraordinary trial which had taken place in Oxfordshire. He had sat upon the bench for many years, and during that time had been associated as a rule with a very considerable majority of Conservative and game-preserving magistrates. But he did not believe there was any of them who would have been guilty of such an extravagant piece of folly as had been committed by those two clergymen in Oxfordshire. He did not know what they could have been thinking of, and he was equally at a loss to know what sort of an adviser they had as their clerk. He did not know how they could have had an officer who would give them such bad guidance. Personally he had strong sympathy for those who had been so treated under the guise of the law. For some fifteen or sixteen years now past he had farmed considerably in this county, and last year he laid before them some statistics as to the wages he had been paying his men. He had been asked to state whether he was able to make it to pay. Well, his answer was this: he had been farming all those years, and it was hardly probable that he would have continued if he had been losing money by it. It was not his custom to go on losing money, nor did he think that was the custom of his countrymen. They would understand by that, that he had not sustained a loss. If he had put his money into some business speculation, he might have made more money than by farming; but he might not have done so. He was satisfied the Union was going on in the right way. Every time he attended its meetings he saw that the organisation was becoming more perfect, and that it was spreading itself more and more throughout the country. He heartily trusted that a branch of the Union might be established in every parish in the country.

The resolution was then put, and carried.

Mr. Jesse Collings then moved the second resolution, which was as follows :—"That in the opinion of this meeting the numerous cases of unfairness and wrong in the administration of justice, which have been brought to light during the agitation of the past year, require immediate attention and remedy at the hands of the Government." He said he disputed one statement made by a previous speaker, to the effect that the farmer was between two millstones. It seemed to him that the labourer was in that position, and was being pretty well ground between the farmer and the landlord. As to the difficulties of the farmer respecting his rent, that was a matter he must settle with his landlord. Perhaps the question might lead to a higher cultivation of the soil; at all events the farmers had the alternative of forming a Union, as they had already done. They might rest assured that the labourers would not refuse to work for them unless they came out of their Union. If the farmers, who were fearfully and wonderfully made, would only consider, they would see that Unions were good for them as well as for the men. Failing relief in any of the quarters indicated, the farmers could do what had been recommended to the labourers—they could emigrate. Or if the conditions of the labourer was to be envied, they could all turn labourers. Referring to the great success which had attended that movement, Mr. Collings said he remembered seeing a paragraph in the papers to this effect last year : "Wait till their next annual meeting, and then you will see them split up into all sorts of divisions." They had not split up yet, and he believed they never would. If there had been a little split amongst them that day, it was only the split of men whose love for the movement was bigger than their patience. It was nothing more than the quarrel of two lovers. The Union had now got its paper. And why should it not, seeing that other movements had their papers. The army had its paper; the farmers had got some two or three papers; the Church had four or five representing its various divisions, but they only wanted one paper. Therefore let them stick to their paper. Of course, in the other papers they would find misrepresentations of all kinds, and these would be repeated in the next day's papers just as they had been before. It would be said that they were in favour of resorting to violence. He was quite sure that untruth would be uttered. He remembered an instance of this perversion of language which occurred at Birmingham several years ago, when they were striving for reform. Speaking at a meeting in the Town Hall, Mr. R. W. Dale said, "If you deny the power to the people—the power of putting wrongs right in a constitutional manner, what is left to them but the pike, the rifle, and the barricade?" That was a very just observation, and it was also a friendly warning. Well, what did the papers say the next morning? Why, what they were saying now. They said that Mr. Dale recommended the people to take the pike and the rifle, and have a revolution in the land. They, however, knew better than utter such sentiments. What they intended was to be free men, notwithstanding the efforts of the farmers, who were doing their best to drive the labourers to desperation. During a period of long-suffering, men were apt to fly to force in order to secure those rights which justices and others failed to give them. But the farmers would not drive the labourers to despair, because hope so abounded in this movement.

He need not mention the oppression referred to in that resolution, for they had lived amongst it all their lives. But the public outside were not so well acquainted with the matter, and therefore they were thankful to those reported cases which brought the facts fully to light. Mr. Collings here proceeded to read the details of several cases heard before the magistrates, in which labourers had been prosecuted for various causes, or had sued their masters. The first was a case in which a labourer was sued for 15s. in lieu of one week's notice. In opening the case for the master, the attorney described the Union as a "detestable association." Well, he maintained that was a libel. The attorney might have said the same thing about the Church. If he had called the Union "a detested association," he (Mr. Collings) could have understood it. In the next case, the attorney said to the man, "Are you a paid agent of the Union?" His answer was, "Yes; I am a paid agent, and receive instructions from Leamington." That was a very proper place to receive instructions from. The next case came from the Highway Board at Faringdon, a body elected for public purposes, and to expend the public money. One of the labourers being called into the room had this question put to him, "You are a foreman?" to which he answered, "Yes." That was a nice thing—a *foreman* on 12s. a week. He wondered they were not ashamed to let the public know they had a foreman working for such wages. The man was then asked if he belonged to the Union, and he said "Yes." The men always said "yes" now. Well, for this alone—being members of the Union—this man and another Unionist were dismissed from their employment, each man having a wife and some half-dozen children to support. Mr. Collings next alluded to the Chipping Norton case, and ironically observed that the women ought not to have put the lives of those two young men in peril by laughing them out of their employment. The farmer, however, saw he had a handle and a hold upon those seventeen women, and brought them before the magistrates. A bold farmer that! Of the sentence passed upon them he had only to say that if it was justice it was not tempered with mercy, and if it was law it was not Gospel. To the labourers he would say, that if they were ever brought before the magistrates, they should try and choose the day, if they could, when there was no clergyman on the bench. Although he had lived all his life in the country, and had had many opportunities for observing the administration of justice, he had never known a lenient sentence, nor anything short of the rigour of the law, come from a clergyman. He hoped they were not going to imitate the conduct of those gentlemen. Mr. Collings next alluded to the Bishop of Gloucester's famous "horse-pond" statement, which he stigmatised as being worse than an incitement of violence, inasmuch as it had not the honesty nor the bravery of that offence. He maintained also that it was a "molestation," perhaps not according to law, but it was so when tested by the spirit of the age which governed men before the law. That remark, too, came from a bishop who was paid a salary of £5,000 per annum. That salary was just £1 14s. 4½d. per hour, and 6s. 6¾d. per minute. Fancy that. And he was not speaking disrespectfully of the bishop. He would never have spoken of him at all if he had not incited to violence. Let them never mind all this oppression, but stand true to themselves, and then they need not fear the result. In conclusion, Mr. Collings dwelt upon the importance of peaceful agitation, and then passed on to the subject of emigration, which he had always opposed, because he thought it was a pity to send the pith of the labouring class out of this country. But he was now beginning to acknowledge the necessity for it. He next proceeded to deal with the question of wages, which had risen 43 per cent. since 1869, and asked how much of it the agricultural labourer had received? The reply was "nothing," to which he answered that it was high time he got something. The truth was, he had been getting worse while everybody else had been getting better. Under these circumstances he could not help recalling to their attention the well-known lines of Goldsmith :—

> "Ill fares the land to hastening ills a prey,
> Where wealth accumulates and men decay."

Well, that had been the state of England for some time. There had been an enormous increase of wealth on the one hand, and no corresponding increase on the other for the labourer. What was the remedy for these things? First of all came the franchise. That lay at the bottom of the whole question. Until they possessed that, they were not true men in the best sense of the term. Over them was a class that would fight against their having the franchise more than against their obtaining an increase of wages. The vote was the only instrument by means of which they could fight their battle successfully. They might even come to move the clerical magistrates off the bench, and then have a word to say about the Church itself. If they had not votes they were not men, for without the franchise there was a degree of serfdom attaching to their position. The vote was the lever by which they moved society, and obtained other things they required. The next thing to go in for was free, compulsory, and secular education. This would enable them to enjoy to the full those rights and privileges which the franchise would secure. Finally, let them persevere, and there was no doubt about their success. Let any one cast his mind back five years, and what was the condition of the agricultural labourer then, in comparison to what it was now? Although not what it ought to be, it was full of hope. He hoped also the men would be backed up by the women, for he had great faith in the women. Above all, let them remain loyal to the Union, and faithful to the woman each one loved.

Mr. Joseph Arch, President of the Union, who was received with an ovation of cheering and applause, said he rose to second a resolution which he could have wished had never been necessary at a meeting like that. When he commenced this agitation, which was just now the talk of the whole country, he certainly had an idea that it would meet with opposition—in fact, he had never known any good movement start but what always had its opponents. Therefore he was not surprised that they should to some extent meet with opposition in their attempts to form a Union for the protection of their interests. But at the same time he certainly must say that he had been surprised at the very injudicious and unfair way in which the labourers had been treated by those who ought to have administered justice. Now, perhaps their opponents, whether on the magisterial bench or amongst the landlords and farmers, thought by these acts of injustice to drive away the Union, and cause the men to despair. In that case, he would tell them most distinctly that — having such a hope in their combination which had accomplished such great things in one year—all the injustice which could be brought to bear upon it would never, he trusted, make it desirable to use anything improper in so noble a cause. He was not going to refer to cases of oppression towards the labourer, because his esteemed friend, Mr. Collings, had given them cases which every sensible man ought to denounce in a land like England. He concurred in the opinion expressed by their friend Mr. Campbell, that the farmers of this country were short-sighted in reference to this movement, and he believed that not only the farmers, but the landlords also, were similarly mistaken with regard to the same matter. He sincerely hoped that, whilst they were so infatuated as to shut their eyes to facts—which, in his humble opinion, did not augur well for the greatness of this country—the Union would, by virtue of its agitation, raise the minds of the public in general, so far as to lead them to see that if the landlords and farmers, by their oppression towards the Union and Union men, jeopardised the interests of England, they would say to these farmers and to those landlords, "We are subjects as well as you, and we mean to be protected against your unjust actions." Now he was almost inclined to think that at the present time the case which had been referred to at Chipping Norton was beginning to make many of the clergy of the country feel themselves to be in an awkward position. He wished to be allowed to express his own views and feelings with regard to the position of a minister of the Gospel. He, for one, respected—and he might say that respect was second to none—an earnest minister of the Gospel of Jesus Christ. Every earnest minister ought to be well cared for by the flock to which he preached and administered spiritual truth. He should be exonerated as much as possible from the cares and anxieties which belonged to the life of the man in business; but whether he belonged to the Church of England or a Nonconformist body, when he left his sacred duties to sit upon the magisterial bench, and after preaching, "Be ye all one, even as Christ your Master is one," supplemented his teaching by sentencing sixteen poor women to gaol, he lowered his office in the face of the world, and rendered his ministry of no effect. When they spoke of Mr. Miall and his supporters in the House, they must not forget the action of those gentlemen at Chipping Norton, which would undoubtedly have a tendency to give him an overwhelming majority the next time he attempted to disestablish the English Church. They had done more by that single act to disestablish the English Church than the Liberation Society had ever done before. There was one case to which he would like to call the attention of that audience. He had no doubt they did want knocking off the bench, and he would tell them how and when they would be knocked off the bench. The agricultural labourers would knock them off the bench when they had got the franchise. He would ask, what were the agricultural labourers to understand by the way in which the law was administered at the present time? They had to submit to the administration of the law. Well, if a farmer could discharge a man at a moment's notice, as had been held by some of the county court judges and magistrates, the labourer should be allowed the same latitude. If that law bound the labourer to serve his employer a week's notice, it ought to bind the employer to give an equal notice in return. But that was not the only mischief. If the farmer discharged a man without notice because he did not want him, the man might seek redress. And if he got redress, what was it? To begin with, it would cost him something to get it, and when obtained it would only be a week's wages. On the other hand, what could the farmer do if the man left him without notice? Why, he had power to send him to prison without the option of a fine. What was wanted was justice, and they were determined to agitate until they got it. He was just about to make an allusion to the highway board at Farringdon, which had discharged several men because of their having joined the Union, and it was to this effect—if this sort of thing were to go on, he should not be surprised if eventually some of the young men present were discharged even for getting married. He was surprised to find that although the Hon. Auberon Herbert had inquired in the House of Commons if the House had considered the discharge of these men, the question had been shelved. Now, he (Mr. Arch) wanted to know if the Government was a Liberal Government or not? He was not going to speak

SPEECH OF MR. ARCH AT CONFERENCE. 179

disrespectfully of the Government, but he must say that we had had too much of that "dodgery," of that obnoxious thing called " Liberalism," which when put to the test proved to be only rank Toryism. Again, he would ask if that Government was practically Liberal which would allow a body of honest labourers that had broken no law, violated no right, injured nobody, to be coerced and oppressed as the labourers had been, drifted out of their employment, to go— God knew where. Such a Government was an anomaly and a monstrosity; and they did not want any more Liberalism of that kind. He would now sit down, as there were several other gentlemen to address them. But he must first make some reference to the tendency of the "justice" dealt out to the agricultural labourers. In consideration of this persistent conduct, it was really wonderful that they had never resorted to violence. That they had not done so was not because they had not had good provocation. He knew there were many in this country so thin-skinned that they had come to regard the labourers as a dangerous class of men. Well, as one of them, he thought he might say that the treatment they had received had proved them to be the most docile, the most willing, and the most forbearing of any class in the country. But he was not going to say to them that night that the labourers were always going to stand such humbugging as they had to suffer. He had been in the town of Buckingham, at a meeting which was frequently interrupted and disturbed by a number of the Yeomanry, whom he called "butterfly shooters." They came to that meeting in the uniform purchased by the ratepayers, to disturb the labourers in the quiet and peaceable discussion of their grievances. Repeatedly he requested them to be quiet, and explained that the object of the meeting was a peaceable one, and what he was going to say was nothing but truth. Well, those men would not desist from interruptions, nor would they allow the business to proceed. Some of the labourers looked at him in a manner which seemed to say, "Shall we put them out?" At last, when it was impossible to bear with the men any longer, he said, "Put them out," and in one minute they were flying across the square, and glad to take refuge in a neighbouring public-house. Now, as he had already said, their opponents must not think that the labourers were going to stand humbugging for ever. If the farmers and others would by oppression and tyranny try to crush the very life out of these men, he would do his best to keep them in order; but if their opponents were determined to crush the very life out of the men, sooner than they should be made door-mats and the slaves of the farmers, he would loose them to a man. As they were all aware, the work of emigration was going on, and that was a matter of importance to the country. Reading history, he found that in the reign of Henry VIII. the landowners of the Isle of Wight thought it was better to rear sheep than to rear men. Well, what was the result of that? The French heard of the depopulation of the country, and came and swept everything before them. In the House of Commons, the other day, a landlord said, "Oh, when the labourers are gone, we'll turn the land into grass as we have done in Ireland." They had been told to reduce the surplus labour, and then matters would be improved. Why, he could take them into many places where they would not find the labourers averaging one to the 100 acres. Mr. Arch then proceeded to deal with the abuse and misrepresentation of the press, and specially named an attack upon himself by one paper which had called him "the Arch-Apostle of Arson." He had never burnt any one's ricks, nor set fire to any person's property. If he had preferred such a charge against the landowners and landlords they would have brought an action against him and he would have been incarcerated in prison; then why should they be allowed to do the same thing against himself with impunity? In one of the local papers he had seen an observation with respect to the evidence he had given in the House of Commons touching the game-laws. The writer said that Mr. Arch had stated before the House of Commons committee that he had known many agricultural labourers given to taking hares and rabbits because they wanted food. He supposed that if the game-laws were abolished these same agricultural labourers would take to sheep stealing and foul stealing for a similar reason. When a man inserted in the public papers such unfair, unjust, and absurd statements they had the right of reply; and if the writer thought fit to box their ears because they spoke strong things,[1] they would box his ears because he wrote like a fool. In conclusion, Mr. Arch referred to the great success which had been achieved by the Union, and expressed a hope that all would stand firmly by the cause, which would eventually triumph over all difficulties.

From some cause or other the agitation of the field labourers has gradually become more political than at first. It has come to be realised by the leaders that no half-and-half measures will meet the necessities of the case. The comments of the London press on the conference proceedings probably contributed somewhat to this result. The "Times," in an article on Mr. Dixon's speech, had the following very suggestive remarks:—

Are we to elevate the British labourer by telling him to be entirely satisfied with the very worst feature of his degradation? Mr. Dixon can hardly intend this, but it is a point on which he has to explain himself much more clearly than he did the other day at Leamington.

And in another place the writer observes :—

We cannot think that the question before us is simply one of labour and wages. It is rather the real and complete elevation of the agricultural labourer to a higher and larger share of our common, and perhaps we may add our Christian, humanity. He is to be made a more independent, more self-governing, more rational, more social personage—in a word, more of a man. He is to be made really a citizen of this great commonwealth, and more worthy of the franchise which one day he is to have.

Mr. Arch and his coadjutors have not been slow to take the hint thus given to them by the leviathan of the press; and hence, probably, the widening of the area of controversy and agitation.

On the 23rd of July, of the same year, Mr. Dixon presented a petition to the House of Commons from eighty thousand farm labourers in favour of household suffrage in counties. On the motion for the second reading of the Household Franchise (Counties) Bill, Mr. Trevelyan proceeded to move it in a very able and effective speech, concluding as follows :—

It is impossible to saddle any individual with the responsibility of a measure which the nature of things is rapidly and irresistibly bringing to the front. We draw a distinction almost unknown in any constitutional country or in our own colonies, which did not exist even here in its present invidious and aggravated form before 1867, between the inhabitants of the towns, and the inhabitants of rural England. We brand our village population as if they were political pagans, just as their class were accounted religious pagans in the days of the Roman Empire. At a time when we must soon be debating questions nearly concerning their welfare, bodily, intellectual, and religious, we cut ourselves off from all acquaintance with their own opinions as to their own affairs, except such as comes to us filtered through the medium of the speeches and resolutions of self-elected politicians, responsible to no colleagues, and to no constituency, and all this we do, not because there is any reason for it in logic, in justice, or in common sense, but because it was so ruled by the wisdom of your ancestors—that is to say, because 400 years ago one of the worst Parliaments that ever sat in this country, robbed the county inhabitants of their votes, on the ground that (to use the very words of the Act), " being people of small substance and no value, they pretended a voice equivalent with the most worthy knights and esquires." There is a story told of Napoleon, who, in the park at Fontainebleau, noticed a sentry walking to and fro in the middle of a grass plot. On inquiry, he ascertained that in the time of Louis XIV. some young trees had been planted there, and a soldier was placed to keep off the cows. The plantation had grown up, had withered, and all traces of it had disappeared, but the sentry walked there still ; and so we keep up a difference between the town and the county franchise, because in 1429 a Parliament of Henry VI. was afraid of our rural population. That fear has altogether passed away. What danger is there for us in giving the franchise to the householders ? They are the heads of families, the industrious, stationary population of the country. We are accustomed to assert that the mass of our people are loyal to the Queen, tender of the rights of property, attached to the institutions under which they have the happiness to live. Well then, by voting for this Bill we shall prove that we believe what we say. By voting against it we shall show that we use this language from the teeth outwards. In the hope that I have said nothing to damage a cause with regard to which my desire is that the day may come quickly for me to deliver it over to more potent and responsible hands, I beg to commit this Bill to the earnest consideration of the House.

Other members carried on the debate.

Mr. O. Morgan, as a county member, said if he had consulted his own convenience, or his own pocket, this would have been the last Bill he would have touched. County constituencies were already unwieldy enough; but inasmuch as this was a change which must come sooner or later, he thought the House might as well make up its mind. There was a time when the peasant belonged to a different class and represented a different interest. Now the urban population had overflowed into the country; and, given the same amount of education, he very much doubted whether the rural population were not quite as much to be trusted as the urban population. No danger was to be anticipated from the change proposed in the Bill; but he did see danger in nine hundred thousand or a million men brooding over real or imaginary grievances, and whose wrongs were aggravated by the thought that they were not to have a voice in redressing them. This question was rapidly coming to the front; and when the next appeal to the country came, it would form one of the most important points raised by the constituencies. The honourable member concluded by seconding the motion.

Mr. T. Collins drew attention to the fate of proposals of this kind, repeatedly made by the honourable member for East Surrey, for counties, and the honourable member for Leeds, for boroughs, and deduced from these instances that no such change could be made, except as part of some great general measure. What was wanted in the House was a number of intelligent and capable men, and not mere delegates to carry out any set of miserable crotchets which ignorant constituencies might wish to see carried out. The honourable member entered in great detail into discussion of propable effects of the change proposed on county representation—a change which he said was of far too great importance to be dealt with by what was at this late period of the session only an abstract resolution. He begged to move the previous question.

Sir J. Kennaway, in seconding the amendment, thought that it would be admitted on all hands that at this period of the session it was mere waste of time to go on with this Bill. There might, perhaps, be some discussion in which members would feel themselves called on to make pledges which there was not the slightest intention to redeem. He denied that the feeling of the country was in favour of the Bill, because working men, who were really representatives of their class, knew very well that they would be welcomed in that House. He did not think the agitation in the rural districts would be permanent, or that there was any real desire to re-open the question of reform. He thought no one could deny that the condition of the agricultural labourer had been much improved of late years. The honourable member seemed to want more steam put on, but he had probably heard the results of putting steam on and sitting on the safety-valve of an engine. Recent legislation had effected such great changes in the direction of extending the suffrage, that they could well afford to wait awhile.

Mr. Cadogan said that he represented a district in which there were thirty-one parishes, thirty of which were purely agricultural, and one alone partook of a manufacturing character, being that of Swindon. No constituency could have left greater liberty of action to their member than his constituency had done, and he totally denied that if the franchise were extended to the labourers, they would be sending up mere delegates to Parliament. The question whether the agricultural labourer ought not to have the same franchise as his neighbours in the towns, was one that required immediate settlement. He contended that he ought to, and that sooner or later he would. He believed that the public activity on this subject would increase instead of dying out, as the Conservatives fondly hoped.

Mr. B. Beach said that the county franchise had always been based upon property. He did not think it would be advantageous to make the county constituency too unwieldy. What was wanted was a careful representation of all interests, but the Bill was only a step in the direction of electoral districts.

Mr. Dixon had rejoiced to hear so many admissions from honourable gentlemen opposite, that the agricultural labourers were as well fitted to exercise the franchise as the artisans in towns. He could assure those honourable members that there was not the slightest jealousy of the counties on the part of the towns. He would also admit that whenever the House approached the question of a fair distribution of political power, a very much larger number of members must be given to the counties. He entirely concurred in what had been said on the other side as to the fitness of the agricultural labourer for the franchise. In this conviction he was strengthened by what he had seen and heard at recent meetings in connection with the new movement, which had originated with the labourers themselves, Mr. Arch himself being an agricultural labourer.

Mr. Newdegate considered that the present discussion was a complete waste of time, because nothing could result from it. He believed the adoption of such a measure as this would tend to universal suffrage, which he should in every way oppose.

Mr. W. E. Forster said he should give his support as an individual to the Bill. It was not his honourable friend's fault that he could not bring it forward at an earlier period of the session. This was a question in which the people of this country took great interest, and especially that part of the community which had no direct representation in the House, namely, the agricultural labourers. He thought, on the whole, the opposition to the Bill had not been as to the principle, but rather to the details, and he congratulated his honourable friend on having introduced it. The principle of the extension of the suffrage to the country householders seemed to be admitted by the honourable member for Warwickshire, though he was rather dolorous about it.

Mr. Newdegate—I don't desire it.

Mr. Forster — Still the honourable member seemed to look upon it as a foregone conclusion, and classed it with one of those subjects on which he uttered forebodings, but seemed to be very comfortable after they passed. He seemed to think it would lead to democracy. As an old household-suffrage man he (Mr. Forster) looked forward with pleasure to the probable extension of the county franchise—nor did he believe it would lead to democracy, or even to manhood suffrage or woman suffrage. The honourable member for Boston complained of piecemeal legislation, but in point of fact, the county franchise was about the only thing which had not been dealt with by piecemeal legislation. The right honourable member for Buckinghamshire, who gave the country household suffrage in the boroughs somewhat unexpectedly, had often comforted his more fearful followers by pointing out to them the limitations that still existed; and the question now was, should that comfort be any longer continued? The right honourable gentleman had never expressed himself of opinion that the agricultural labourers were not as well fitted a people as the same class in towns to have the franchise, but he (Mr. Forster) doubted very much whether, taking the average, the labourers were not quite as well able to exercise the franchise as mechanics. Day by day the agricultural labourers were taking more and more interest in political matters, owing to the recent movement in the agricultural districts, and they had shown that they had as much interest in obtaining the franchise as the people in the towns had. He was not prepared to say that at first the Conservatives would not gain a little by it, but that was not a consideration which would deter him from extending the franchise, while it would be a ground for the support of Conservative members. There was no reason whatever, why a man in a manufacturing village in Lancashire should not have a vote just as well as a man in a borough. In what were termed county boroughs, the anomalies were becoming more and more startling. The House surely could not adopt the statement that, to extend the county franchise would necessarily lead to manhood suffrage. They must all feel hopeful of the future of their country when they saw such a large portion of the community taking so deep an interest in the matter. It deeply concerned that House and the country that they should have found the agricultural population—when they were awakening to their rights and engaging in a dispute with their employers—conducting themselves with such moderation as to afford an example rather than a warning to citizens. The fact that those persons who had not votes were uniting together was the very reason why the House ought seriously to consider how long they should be without. A petition signed by 82,000 agricultural labourers was not one which the House of Commons would ignore, and it was very possible that some discussion might arise in Parliament itself upon this question, and perhaps such discussion might be followed by legislation. He was of opinion that it was quite indefensible that this large class should feel that they had little, or no direct voice, in the government of the country. They could not forget that the county members represented not the employed, but the employers, and that must needs be a question for the House to consider when questions respecting employers and employed came before them. Those were the grounds on which he felt able to vote for the Bill ; and he might add, that his right honourable friend the Prime Minister, who was confined to his room by indisposition, had desired him to state that while the Government expressed no opinion, and made no recommendation on this matter, and while he regretted that it should have been so long postponed, as now to present the aspect of an abstract resolution, yet he (Mr. Gladstone) desired him to state that he retained the opinion which he had more than once uttered, that the extension of the franchise in counties was just and politic in itself, and that it could not long be avoided.

Lord J. Manners (to whom Mr. Fawcett gave way) considered that it was very improper of the Prime Minister, whose absence from the cause stated they all regretted, to send down such a message as that just delivered, and that it was taking an unfair advantage of the forms of the House. The right honourable gentleman told them that this was an open question. Just let the House imagine a question of such magnitude, and involving such important issues, being thrown, in the dog days, on the table of the House of Commons as an open question. They were not told what the opinion of the Cabinet was, and it would go out to the country while her Majesty's Government carefully abstained from giving a collective opinion, as the responsible advisers of the Crown, they are willing that this question should be trailed during the recess as one which received the individual support of the Prime Minister and the Vice-

President of the Council of Education. When they were told by the Vice-President that momentous issues were bound up with the second reading of the Bill, which they need not now discuss, but that they might leave it to time and chance to settle how many seats, as the necessary consequence of its passing, should be taken away from the boroughs and given to the counties, he thought the House should refuse to give its assent to a Bill which dealt with a fragment only of this great question.

Mr. Fawcett said, that after the speech of the Vice-President· of the Council, and the message sent to them by the Prime Minister, it could not be doubted that this Bill was virtually taken out of the hands of his honourable friend, and was now a settled part of the policy of the Government. It was absolutely impossible for the Prime Minister and one of his most influential members of the Cabinet to vote on such a question as this as ordinary members of Parliament. They voted for the Bill as members of the Government, and it henceforward became undoubtedly a Government measure. Therefore, what he would impress particularly on the House was, that it was no longer the assertion of an abstract principle, but the beginning of another great measure of representative reform ; and the question they had to consider was this—were they going to sanction a great extension of the suffrage again, without having any definite statement from the Government of what should be the principle which should regulate the distribution of political power? Now, although he was as ardent as any man could be, in favour of the extension of the suffrage, yet when a bill with that object was introduced by the Government he would not vote for it unless he knew what were the principles Government were going to adopt with regard to the redistribution of political power in this country. They might have the most democratic suffrage in the world, and yet, if they did not take security that minorities should be represented, that democratic suffrage, by concentrating unchecked power in the hands of the majority, might lead to the worst of all kinds of oligarchy. The principle which should regulate them in trying to improve the representation of the people should be, on the one hand, to enlist as voters every single person who was not disqualified to vote, and at the same time to take care that that House should not simply represent local majorities, but should be truly a great national assembly, in which every opinion should, as far as possible, be represented by its ablest and most independent advocates.

Mr. Bruce said his own individual opinions on this question had been so fully represented by the speech of his right honourable friend the Vice-President, that he should not have risen on the present occasion if it had not been for the remarks that had been made in reference to the Government. The noble lord opposite had found fault with the course taken by the Prime Minister, but he (Mr. Bruce) thought the House would feel that it was only natural and respectful to the House that, when a discussion of such importance was proceeding, his right honourable friend should explain the cause of his absence, and should add to that, that the opinions he now entertained were those which he had always expressed in favour of the principles embodied in the Bill of the honourable member for the Border Burghs. Any one who heard the noble lord would have supposed that this was an occasion which had been seized by his right honourable friend at the head of the Government for making a political manifesto of a new and startling kind. But was there one honourable gentleman who cheered that statement who did not know perfectly well that on more than one occasion his right honourable friend had given expressions to those opinions? Then the honourable member for Brighton stated that the Bill having received the support of the Prime Minister and the Vice-President it must be considered to have passed into the hands of the Government. He (Mr. Bruce) denied the right of any honourable member, however respectable, to come forward on any occasion he pleased, and challenge the opinion of the Government as a Government on any particular question. This was a question which required great study and careful thought, and any Government which, without having given it that study and thought which were necessary, would express an united opinion, would be rash indeed. There was, however, no real force in the argument of the honourable member for Brighton, for the instances were not few in which individual members of the Government had expressed their opinions in reference to particular subjects without in any way committing their colleagues.

Mr. Scourfield asked the House to pause, before it sanctioned the principle of identity of suffrage, which was described by the Secretary of State for War in 1859 as "laying the axe at the root of our national liberties, and destructive of the British Constitution."

Mr. J. G. Talbot characterised the message of the Prime Minister as a political manifesto to the country.

Henceforth the franchise question was a plank in the Union platform. At every village meeting the coming event of political enfranchisement was an important part of the discourse. And the instinct of the labourers which leads them to make much of the prospective boon, is undoubtedly a true one. The day of their elevation to the rights of citizenship will certainly hasten on the day of their social elevation. It will be impossible to keep half a million of parliamentary voters in the state of semi-pauperism, which the agricultural labourers are, for the most part, in to-day. Invested with the dignity of political manhood, they will soon cease to be the mere burden and difficulty of the poor-law board.

It is not the least significant feature of the labourers' movement, that already it has told upon the pauperism of the rural districts. The bare assertion of their manhood, to which the young and energetic organisation can hardly, at present, be said to have much more than attained, has gone far towards making the workhouse and the relieving officer an impertinence and an insult. In a journal published in the very heart of a great agricultural district—one of those Cricklade "hundreds" of which Mr. Cadoganwas for years the worthy Liberal representative, I find the following statement :—

At the weekly meeting of the Highworth and Swindon board of guardians, on Wednesday, April 16th, the clerk reported that the guardians had in the hands of their treasurer £3,981. Every half-year, contribution orders are made upon the various parishes in the Union to provide funds for the relief of pauperism, &c. Although these orders have not of late exceeded in amount any former ones, the guardians, after meeting the liabilities consequent upon the pauperism of the winter quarter, have the above-mentioned balance left in hand and uncalled for! This fact cannot in any way be attributed to a favourable winter, for it is well known that the past season has been unusually unfavourable for agricultural operations. But it may with great safety be alleged, that the slight increase agricultural labourers have been enabled to obtain in their wages has saved so many families from being driven into pauperism, that the public rates are affected in this one Union alone to the extent above indicated. Of course it is only in the relief given to paupers that any saving can have been effected, for the establishment, county, and other charges have remained unaffected. We believe we are correct in saying that the balance now in hand would be ample for the relief of the poor in the Union for the next twelve months, without another penny being raised. Nor is this state of things confined to one Union. In the adjoining Union— the Cricklade and Wootton Bassett Union—the guardians are advertising for two general servants to do the necessary work about the workhouse at Purton, there being no inmates to do it.—*Swindon Advertiser.*

The returns for the agricultural county of Berkshire, are even more conclusive testimony on the subject.

POOR RELIEF IN BERKSHIRE.

The following is Mr. Henley's return of the cost of poor-law relief in the county of Berks for the last four years:—

	Population, 1871.	Total number relieved 1st January, 1873.	Being 1 in
Abingdon	21,552	1,343	16
Bradfield	15,852	987	16
Cookham	14,875	770	19
Easthampstead	10,630	466	23
Farringdon	15,091	431	35
Hungerford	19,347	1,096	17
Newbury	20,640	1,381	15
Reading	33,330	1,169	29
Wallingford	14,641	877	17
Wantage	17,367	1,014	17
Windsor	26,805	760	35
Wokingham	16,191	907	18
Total	226,322	11,201	20

IN-MAINTENANCE.

	1870. £	1871. £	1872. £	1873. £
Abingdon	1,849	1,807	1,670	1,823
Bradfield	1,929	2,152	2,210	1,792
Cookham	1,695	1,740	1,848	1,805
Easthampstead	829	883	906	1,021
Farringdon	1,114	1,130	1,082	1,095
Hungerford	1,259	1,349	1,280	1,275
Newbury	1,947	1,900	2,000	1,910
Reading	3,778	3,921	4,314	3,288
Wallingford	1,210	1,290	1,338	1,424
Wantage	819	940	784	764
Windsor	2,677	2,930	2,644	2,534
Wokingham	1,685	1,781	1,674	1,261
Total	20,791	21,823	21,750	19,992

OUT-RELIEF.

	1870. £	1871. £	1872. £	1873. £
Abingdon	6,547	6,898	5,607	5,043
Bradfield	4,373	4,632	4,384	3,887
Cookham	2,557	2,939	2,893	2,927
Easthampstead	1,400	1,632	1,481	1,258
Farringdon	1,401	1,383	1,340	1,297
Hungerford	3,801	3,973	4,097	4,014
Newbury	6,629	6,819	6,861	6,119
Reading	3,296	3,621	3,226	2,780
Wallingford	5,405	5,515	4,596	4,203
Wantage	4,502	4,650	4,141	3,895
Windsor	1,915	2,322	2,318	2,325
Wokingham	5,508	5,417	4,290	3,605
Total	47,334	49,801	45,234	41,353

These returns show that the Farringdon district of 15,091 inhabitants has only 431 paupers, and spends £1,095 for in-maintenance, and £1,297 for out-relief, which is lower than any other district proportionately. And side by side with this fact stands this one, that in no other district has "Unionism" taken such deep root. The West Berks' branch, which commenced in 1872 with ten members, now numbers more than three thousand. Hundreds of pounds have been remitted by the members to Leamington, and the ratepayers have been saved a corresponding sum.

The next and last incident of the agrarian agitation to which I intend referring, is the journey of Mr. Arch to Canada. While remote contingencies, such as a readjustment of the land laws, and the possession of a vote, were engaging the attention of the members of the National Union, and while their organ was weekly pouring forth volumes of most valuable instruction upon the somewhat bewildered understandings of its readers, the pressing necessities of their every-day life were calling aloud for present and practical measures of relief. In a word, something was wanted to be done, and that quickly. Philosophers might demonstrate that by other arrangements of the land of England, and an alteration of its laws, things might be made much better for the agricultural labourer, but meanwhile that same valuable individual was starving on ten or twelve shillings a week, and he was growing very tired of the process.

Something must be done, and practical men were not slow in seeing that that something was EMIGRATION. From all parts of the world pressing and urgent demands were made for agricultural labourers. In every colony the cry was heard for field-workers—men with sinew and bone to till the earth, and gather in the rich bounty of its exhaustless treasury.

Hence, the somewhat reluctant, but at length earnest appeal of the labourers to their leader to go forth and see what promise of deliverance the "greater Britain" would afford. Canada, as the nearest and most accessible of our colonies, was fixed upon as the one first to be visited. Of this mission, the series of letters which accompany this sketch of the history of the "Revolt of the Field," will afford some little information. They were addressed originally to the "Daily News," with a view to the world-wide diffusion of reliable information respecting our north-west dependency. A considerable amount of controversy having arisen out of some of the statements contained therein, it has been again and again suggested to the writer to republish them in a collected form. Having, with some considerable reluctance, at length yielded to the request, he has, at the recommendation of his publishers, prepared this brief history of the great movement from its rise to the time of Mr. Arch's departure for Canada on the 28th August, 1873. It is not, of course, anything like a complete history of the movement; that is reserved for some abler hand at some later date. Such as it is, however, it is sent forth, in the earnest hope that it may contribute to the removal of a portion of that distrust and opposition which have marked its progress hitherto, and which have been the only causes of anything bordering on anxiety respecting its future, which its friends have experienced. In itself an unmistakably good thing, as a systematic effort on the part of a too much neglected and sadly underpaid class of workers to raise themselves in the social scale, it has sometimes been feared lest through an obstinate resistance of their reasonable demands, and that oppression which drives wise men mad, resort should be had to a violent settlement of the question. Thanks, however, to the high moral and religious character of the leading spirits of the revolt, and the presence in their council chamber of men who, to a sincere respect for the honest toiler, united the moderation which comes, sometimes at any rate — we cannot say always — of superior culture and higher social position, no real cause for anxiety has ever yet been experienced. A singular moderation has characterised the movement from the first. No instance of a vindictive spirit has stained its history. Men have been again and again turned out of home and work for no other cause than identification with the Union. Justices of the peace, lay and clerical, have dealt out to them a merciless justice, and in not a few cases even the very letter of the law, as well as its spirit, has been nearly, if not quite violated, at the bidding of a relentless detestation of the movement. Refused public halls and other places of resort, they have had to do costly battle for even the right to meet in the public market-place, or on the village green. The sad necessity for parish relief has been made additionally degrading by the strong anti-union feeling of its administrators. Workmen of a highway board composed of landlords and farmers have been summarily dismissed from their employment because they had named the name of Arch. A thousand indignities have been heaped upon both the Union and its disciples, but there has been no wilful retaliation, no agrarian outrage, no loss brought home to an employer's door. And there can be no doubt that to this pleasing fact the marvellous success of the cause is in no small degree attributable. An unwilling homage, like that which has latterly been paid it by the leviathan of the press, has thereby been extorted, while, as the following correspondence will show, it has secured to it the warmest admiration of our fellow-subjects across the Atlantic.

The name of Joseph Arch is fragrant as incense all over the American as well as North American continent. Not as the instigator of a lawless conspiracy is he known throughout the length and breadth of that mighty country, but as the leader of a loyal and law-abiding host of worthy citizens, whose only crime is their poverty, and whose only demand is the right to live by their toil.

May the same moderation characterise the agitation till its work is done! Then will the future historian have to record among the triumphs of the nineteenth century—among those mighty though bloodless revolutions which the religion of Jesus Christ was destined to effect—that of the "Revolt of the Field," which is the designation I have chosen for the work of the National Agricultural Labourers' Union of England.

WITH MR. ARCH IN CANADA.

QUEBEC, *Sept.* 9.

AS some misapprehension seems to exist as to the mission of Mr. Arch in America, it may be as well to preface my correspondence with a brief sketch of its origin. For some time it had been growing apparent to the executive committee of the Labourers' Union that such measures as the migration of labour from one part of the country to another, would but very inadequately meet the necessities of their case. Mr. Arch and other prominent leaders of the movement had hitherto steadily set their faces against emigration. Sympathising, perhaps, too largely with that love for the old home which has kept so many thousands from pushing their fortunes in other parts of the world, these men have contended that their fellow-labourers have a legitimate claim upon the owners of the land which their toil enriched, for comfortable and adequate support. "Why," they have said, "should we be driven beyond the seas to obtain a livelihood, when millions of acres of land are lying uncultivated in England?" But the emphatic *non possumus* of the landlords to this somewhat unusual reasoning, combined with the difficulty of obtaining increased wages and better house accommodation, has at length forced attention to the practical measure of wholesale emigration. This position reached, the steps to the present mission were few and short. If it was best for the labourers to go elsewhere, whither should they go? A hundred eager claimants soon appeared. Each colony and each foreign Government had its agents in the labour market bidding against one another for the much coveted article—for the English agricultural labourer is known far and wide as the best thing of the sort on the face of the globe. But the executive committee, composed as it is of farm labourers alone, or at any rate of men who were such but yesterday, received the overtures of these agents with much caution. The lamentable breakdown of the Brazilian scheme had done not a little to shake their faith in emigration. Hence the mission of Mr. Arch. The offer of a member of the consultative committee to accompany him being cordially accepted, the deputation sailed in the *Caspian* on Thursday, the 28th August, for Quebec. The weather, which at first worked sad havoc among the passengers, improved after the first two or three days. As all travellers know, Messrs. Allan have contributed their full share towards that perfection of oceanic transit accommodation which makes an Atlantic trip little more than a delightful recreation. On the 4th September we had the double gratification of seeing in the distance sundry monster icebergs and the hills of Newfoundland. The former phenomena were more interesting to the passengers than to the captain. The caution and watchfulness exhibited in presence of these terrific ice mountains are beyond all praise. We had a number of Canadians on board, and it was pleasing to us Englishmen to listen to their enthusiastic loyalty.

Another peculiarity of these prosperous colonists is equally striking and satisfactory—their profound satisfaction with their union with Great Britain. "I suppose annexation to the States will be your final destiny," I said tentatively to a group in the smoking-room. "Never!" was the unanimous and emphatic reply.

On Sunday, the 7th inst., we steamed into the magnificent harbour of Quebec. On landing, a gentleman connected with the Emigration Department soon found us out; and after our luggage had been passed, he took us over the extensive emigration barracks which the Dominion Government has erected. Mr. Arch was much gratified by the completeness and variety of accommodation provided. A thousand emigrants could there find a temporary home free of charge. The women have, in a lofty wing of the building, an admirable laundry, where all the family linen could be renovated. Lavatories for both sexes also are provided, and ample cooking accommodation. Upstairs large rooms are fitted with sloping benches for sleeping purposes. Driving out for some miles into the country both yesterday and to-day, we had an opportunity of seeing some of the Canadian small holdings. As the settlers were mostly French, the evidence was not of much value as respects the English labourer. A careless, slovenly style of farming seemed to be the rule everywhere. Knowing how much better his English brothers would improve such advantages, Mr. Arch seemed to see in those miles of small farms, with their brightly coloured and tasteful little homes, something like a realisation of his fondly-cherished hopes.

From an interview we had with the Deputy Minister of Public Works here, it appears that the Government of Canada is engaged in maturing an emigration scheme which embraces most of the points deemed essential by Mr. Arch. I am not at liberty to divulge the scheme, as it is at present but very imperfectly developed. Suffice it to say that it is contemplated to supplement the free grants of land with some provision for the immediate starting in life of the settler who, like our agricultural labourers, has no capital. A rough home will be built for him ; seed will be supplied ; a portion of his land cleared. The cost will be repaid after, say, the third year, by annual instalments, ranging over from five to ten years.

Here we have all that reasonable men could desire, and should the results of Mr. Arch's personal investigation of the grants, and contact with those who have already availed themselves of them, be satisfactory, a stream of emigration will probably flow out westward next spring which, for good or bad, will exert a considerable influence on British agriculture.

To-morrow we start for the "bush," and I shall hope to furnish you by the next mail with an account of the English farm labourer at work in his new home.

Mr. Arthur Clayden, who has accompanied Mr. Joseph Arch on his mission to Canada, writes from Quebec as follows to the "Birmingham Daily Post":

On Wednesday, the 10th inst., a message reached us from his Excellency the Governor-General of Canada (Lord Dufferin), requesting us to call upon him. On arriving at the Citadel, where his lordship resides during his visits to the city, we were speedily ushered into his presence. He received us with the frankness and courtesy of the true English gentleman. Taking his seat opposite to us, he soon plunged into the very heart of the subject. Happily, the man who had passed so well through the ordeal of the Game Laws' Special Committee of the House of Commons was quite equal to the occasion. Clearly, intelligibly, and most forcibly did Mr. Arch put before his Excellency his great life mission, and the object of his travels westward. And with equal intelligence and clearness, and with considerable sympathy, was the story received and apprehended. After an hour's interview we left, and on arriving at our hotel we found invitations awaiting us to dine with his Excellency and his lady in the evening. On arriving there, we found a brilliant assemblage congregated, and a very pleasant evening was spent. Several eminent men were present, who, during the evening, took opportunities of conversing with Mr. Arch. Having another engagement, we requested permission to withdraw at an early hour, and his Excellency at once came forward, shook us very warmly by the hand, and bade us God-speed on our journey, at the same time promising to write a letter of introduction for us to his subordinates throughout the Dominion. To-day we are off to the "bush," to see how the toilers fare, in the apparently "good land."

SHERBROOK, *Sept.* 15.

On Thursday, armed with the letter which his Excellency had kindly promised us on the previous day, we started for this thriving part of what are called the Eastern townships. Our first impression of Canadian railways was on the whole favourable, though I should much prefer a two hundred mile ride on our "Great Western" to a hundred mile ride on this "Grand Trunk." There is so much shaking that an unpleasant impression is produced that a screw is loose somewhere. These light wooden bridges which span the various rivers *en route* may be safe enough, but they don't look so. An Englishman, accustomed to the solidity and elaboration of workmanship of his own costly lines, finds it difficult to believe that these simple steel-lined tracks are all right. The bridge, however, carried us safely over it on Thursday, and therefore we are bound to speak well of it. The scenery through which we rode was exquisite. The authorities of Quebec had deputed a gentleman well acquainted with the district to accompany us. On Friday, escorted by this gentleman and another Government official established here, we started to spy out the land. By Saturday night we had gone over some eighty or one hundred miles. Our first stage was forty miles by rail to a village named Stanstead, from whence we drove back to Sherbrook. This drive took us through several villages and past numbers of farms. I am sorry to say our investigations were not satisfactory. If the farmers whom we met last week are a fair specimen of the Lower Canada farmers, I would earnestly dissuade the English labourers from leaving their present masters to come out and serve under them.

Toilworn, narrow-minded, and apparently without one other idea than that of how much work they can get out of a man for the dollars they must pay him, I know of no agriculturists in England whom I would not elect to serve under in preference to them. "What are your hours?" we said to one of the farmers who intimated his desire to have an English labourer sent out to him. "From sunrise to sunset during five months, and from six to six during the rest," was his reply. "Then all I can say to you," replied the outspoken Warwickshire man, "is that I wish you may get him." "But our pay," continued the farmer; "consider how good it is—a dollar and a quarter a day, with board and lodging." "Can't help it," responded Mr. Arch; "what you want is a slave, and Britons never will be slaves." The dried-up, labour-starved owner of hundreds of broad acres seemed as much nonplussed as English employers have been by Mr. Arch's strong utterances. The truth is, these men have led such a hard, tough life, that they are not likely to be very considerate to others. Of course there are many exceptions, but this was the rule with these self-made men, as far as we saw them. On my expressing astonishment at the absence of labourers from the farms—for, driving all day through a farming district, I saw no men at work anywhere, except here and there one, whom our guide assured us was either the farmer or his son—a farmer with whom we stayed to converse assured us that they got on very tolerably. "Yonder," said he, "is a farmer who is worth 6,000 dollars and a farm of 300 or 400 acres, and all his ordinary help is one young fellow whom you see now with him." And, sure enough, as we drove past, there was the tough old fellow slaving away with his rake among the barley, and close at hand was the one farm hand. The comfortable, jolly-faced farmers of Old England need not grudge these Canadian farmers their rent-free domains. Verily there are worse things than rent audits. I have seen more haggard-faced farmers since I have been in Canada than I have met during a forty years' residence in rural districts at home. And never have I seen during the same period such miserable-looking, lank and hopeless labourers as the few whom I have seen in the service of these terrible task-masters.

I am afraid this testimony will spread something like dismay among the Unionists who are looking to emigration for help from the pressure of their present circumstances. Let it be clearly understood that these strictures do not apply to Canada as the home of the English labourer. If, as I am sanguine will be the case, the Minister of Agriculture at the seat of Government should fall in with some suggestions which Mr. Arch will submit to him, a hope of a very bright and inspiriting character may yet arise for the English labourer in this western colony. But unless, as may possibly be the case, our investigations in Upper Canada should be more favourable than those which we have gone into in this lower province, Mr. Arch will certainly not recommend his clients to exchange their masters of the Old World for those of the New.

One of the interesting features of this rising and most picturesque town is a large cloth manufactory, which Scottish enterprise has established. We dined to-day with the able head of the factory. He has some five hundred hands at work, and the business is growing every year. I am struck with the fine openings presented on every hand for young farmers with a moderate capital. I am convinced that our

English farming is the one thing needed to develop the resources of the country. The bad farming of the men now in possession is its greatest misfortune. One gentleman—I think, or at any rate I hope, an Englishman—has demonstrated in a most remarkable manner what may be done in the way of farming upon this fertile continent. We drove near his estate yesterday, and but for his unfortunate absence from home we should have called and seen his farm. He has within the last six months sold ten head of cattle, reared by himself, for no less a sum than ten thousand guineas. One splendid animal has gone to England, purchased for the sum of three thousand guineas.

We started yesterday on a forty mile drive to a place called Scott's town, where a few enterprising Scotchmen have formed a sort of colonisation society. Some thousands of acres of land have been purchased, and already over a hundred hardy Highlanders and others are there hewing down timber and carving out fortunes. We found an exceedingly interesting community, and the spirit of enterprise displays itself in a 60-horse power engine hard at work driving a saw through the timber as it is felled, and an excellent dam which is being built across the river. By-and-bye water will have to do what steam is now doing, and the surrounding forests will be rapidly converted into lumber for house building and a thousand other purposes. The presiding genius of the colony is Mr. John Scott, a Glasgow man, I believe. He appears to be just the man for his work, and has little doubt of having a thousand families established round his mills before five years are gone. I think he is not more sanguine than the appearance justifies. His scheme is very much like that which Mr. Arch will submit to the authorities at Ottawa next week, as the only one which will, in his judgment, meet the case of English farm labourers. Mr. Scott will build a cottage for each family, and grant an uncleared plot of land. The timber on this land will be purchased of the men by Mr. Scott, so that the settler will not only be getting his land cleared for cultivation, but will be getting a supply of ready money wherewith to live in the meanwhile. The cottages are substantial erections, and appear very comfortable. All the folks seemed happy and hopeful.

The more we see of the country —and I need scarcely say these long drives into the very heart of it give us the best possible means of judging—the more are we impressed with its immense importance for emigration purposes. It would really seem as if half Europe might be gathered into its capacious bosom. There is plainly nothing requisite for man that may not be grown upon its rich and fertile soil. It appears to me that the one grand requisite for the development of its resources is a wholesale importation of Scotch or English agriculturists. Frenchmen make capital cooks— blessings on their skill; but of their farming I am sorry to say that to an Englishman's eye it is as bad as bad can be.

NIAGARA FALLS, *Sept.* 28.

The incessant occupations of our present life leave little time for correspondence. I have brought my table out on the broad verandah which encircles our hotel, and sit down in view of the Falls, of which I will not attempt a new description, to report progress. After leaving what are called the Eastern townships, in the province of Quebec, we reached Montreal on the 16th, and received the hearty welcome and hospitality which have been accorded us from the moment of our landing at Quebec. Government officials awaited us, to see that we were well cared for, and a programme was soon drawn up for future operations. Next day, accompanied by the proprietor of one of the leading journals, we visited the Provincial Agricultural Show, and were introduced to a number of the agriculturists of the province. One gentleman whom we there met, has done perhaps more than any one else to develop the capabilities of the colony. From being an importer of prize beasts from England, he has come to be an exporter; and he told me that he had, within a short period, sent off to Europe ten animals, whose united value amounted to over ten thousand guineas. He has thus demonstrated what can be done in Canada, and certainly the specimens of shorthorns which we saw at Montreal justified any amount of confident anticipation. The show of horses, too, was remarkably satisfactory, and although the sheep were few, there were pens of Southdowns that my old friend the late Mr. John King Tombs would have been glad to number among his fine flocks in Gloucestershire. The roots and cereals were only such as the practical agriculturist, Mr. Arch, had long been convinced might be grown to perfection in these fertile plains. The only reason why so few roots are grown is the universal hindrance — the scarcity of labour. But when the condition of adequate capital, and what capital can command—adequate labour—is complied with, corresponding and most satisfactory results are reached. This ready response of the prolific soil to proper culture has been again and again proved to us. It was only a day or two ago that we spent an hour or two with a gentleman who owns one of the best farms in Canada. He told us that he had grown thirty tons of Indian corn, and nearly as much of root crops to the acre. Indeed, he said he could get off one acre enough to keep a cow the whole year. And this is the land about which one of your contemporaries warned Mr. Arch against inviting his fellow-labourers to, lest they should be starved or eaten up by horned beetles and grasshoppers. Few things have impressed us more forcibly while travelling through Canada than the lamentable ignorance of Englishmen generally respecting the true character and resources of this splendid colony. To think that within a fortnight's distance of England all the unknown and incalculable wealth of this magnificent country should remain comparatively unsought after is positively overwhelming. What are all those hundreds of young farmers thinking about who, with utterly inadequate capital, are fretting against the social bars of their English homes and English circumstances, that they do not come over here and pick up the one and two hundred acre well-cleared farms which lie scattered over the various provinces? In each of these farms is a mine of wealth, and it only needs strong arms and clear heads to develop it.

At Ottawa, the seat of Government, we had several important interviews with the heads of departments. In Mr. Lowe, the Secretary of the Agricultural Department, we found, what we had already found in his chief, the Hon. Mr. Pope, a shrewd, practical, and exceedingly intelligent public officer, who warmly supported Mr. Arch's mission. Nor was the reception accorded us by the Prime Minister of the Dominion, Sir John A. Macdonald, one whit less cordial and satisfactory. He dwelt principally on the vast openings for European emigration in the new province of Manitoba. "There," said this statesman—who, by the way, is wonderfully like Mr. Disraeli in personal appearance—"is a country just opening up, and which a vast railway will speedily traverse, containing untold millions of acres of the finest land in the world, and to every settler we now offer one hundred and sixty acres free of all cost!" One hundred and sixty acres! a decent freehold estate, and to be had for the asking!

Reaching Toronto on the 25th, we found the welcome information awaiting us that the Government of the province would consider us as its guests during our stay in Ontario. Apartments had been secured for us at the Queen's Hotel, one of the largest on the continent, and Lieutenant-Colonel

Denison was deputed to act as our guide throughout the province. When on the next day we visited the Ministry at the Government buildings, we found our highest anticipations as to the attitude of the authorities more than realised. The Attorney-General and Premier of the province, Mr. Mowatt, entered warmly into Mr. Arch's projects, and saw in him at once a very valuable coadjutor. After taking a drive round the city during the day, and inspecting its principal buildings, which, I may remark in passing, are far more massive than might have been expected from a young and struggling province, we met at the Government buildings some of the chief men of the province to dinner. In addition to the ministers, we had the honour of meeting Mr. Mackenzie, the leader of the Opposition in Parliament, and various other Canadian celebrities. Nothing could exceed the universal good feeling, and all appeared in perfect unison on the question of emigration. It was the common admission that the English farm labourer was the one want of Canada, and no one seemed to express a doubt as to the universal willingness there would be to pay almost any reasonable sum to secure him.

The issue of our visit will be a systematic co-operation on the part of the Dominion Government with the Leamington Union, to ensure a perennial stream of first-class emigration. A registry will be kept here of wants, and a descriptive list of such wants will be sent to the office of the Union, and from thence distributed, through its complete and efficient agencies, all over the rural districts. So in all human probability will the great problem of the agricultural labourers' position in England be henceforth solved. If the home employer cannot really afford to pay proper remunerative wages for the services of the men, a clear and straight pathway will be opened before them to a land where those services will be adequately rewarded.

The following extract from my diary may be interesting, as giving the impression produced on the mind by a first view of the world's greatest wonder:—

October 27th.—To-day will be memorable on account of my first view of Niagara. We started at seven o'clock by steamer from Toronto, across Lake Ontario. The day was gloriously fine, the sun shining in a cloudless sky. Beautifully our vessel glided over the rippling waters, and in about two hours we entered the Niagara river. Here we exchanged the steamer for the rail, and along this we rode by the river's side towards the Falls. The scenery was all that could be desired. The steep banks of the picturesque river were clothed with forest verdure—trees of all kinds growing down to the water's edge, reminding me very much of the Leigh woods on the Avon at Clifton. Indeed, if that river were only some twenty times as wide as it is, and about ten times as rapid, it would very much resemble the Niagara river. As we drew nearer the Falls, the river became more impetuous, and the scenery altogether wilder and more impressive. On our way we passed General Brock's monument—a huge stone erection two hundred feet high, raised on a sort of natural plateau on the opposite side of the river. This brave soldier fell there on the 13th October, 1812, while defending Canada from an American invasion. Soon the first glimpse of the Falls was caught. Looking eagerly ahead through the car windows, I saw a cloud of white mist. I knew what it meant, and began to experience that strange sensation which, I suppose, all highly nervous temperaments are subject to on the eve of an important episode in their career. On went our prosaic engine, and soon the stupendous sight burst upon us. There was the unique spectacle right before me—a vast river falling over a precipice one hundred and fifty-eight feet deep. The sun shone brilliantly on the dazzling spectacle, and in mute astonishment I gazed on what myriads have gazed on, and yet no one has described. It is simply indescribable. The Falls on the American side are not so imposing as the Horse-shoe Falls on the Canadian territory. We soon exchanged the cars for another conveyance, leaving our luggage to find its way to the Clifton House Hotel, while we "did" the Falls. After visiting all the points of interest, I had what I must call the grand adventure of my tour, if not of my life. The Cave of the Winds is familiar enough to all Canadian tourists: I will help those who know it not to comprehend somewhat of its romantic nature. My companions deeming discretion the better part of valour, left me to do this part of the programme alone. Entering a wooden shanty just over the Falls, I had first to exchange my clothing for something more waterproof. I then followed a guide down some hundred and thirty steps on to the rocks below. Over these we clambered, amidst the blinding spray from the descending torrent, till a wooden bridge was reached. "Now turn round and look up, sir." I did so, and was appalled. Right in front of me, high up in the air, was a river falling over as if it must wash me and the puny erection on which I stood away like a feather. But no, right down into the abyss eight feet in front of me it falls, and save a baptism of spray, I am safe and unharmed as my companions on yonder eminence. On went my guide through the deafening noise and confusion. I began to question the wisdom of my adventure. It seemed like an invasion of nature's privacies to be there. It was the very fastness of the romantic. O that overhanging torrent! will it not deviate a few feet and wash the brace of inquisitive mortals into the foaming rapids beyond? But nature is not vindictive, and we were spared. "Forward," cried my guide. Well, thought I, how much nearer death? Climbing over the rocks, keeping up as best I could with my agile companion, who like a chamois leaped from point to point—apparently oblivious of the fact that I had not been alike accustomed to danger—until all sense of it was lost. "Hold!" I at length cried out to the remorseless fellow, whose outline I but dimly discerned through the cloud of mist. Vain appeal! A discharge of artillery would scarcely have been heard, I suppose. Soon our vantage point is reached. We are in the Cave of the Winds. Behind us, reaching up some eighty or a hundred feet, was a hollowed rock, and over the projecting ledge came the magnificent torrent in an exquisite curve. Imagination could not conceive any position more romantic. There was the descending water right before us, and pictured on the clouds of spray caused by its descent was a resplendent rainbow—not a mere arc as we behold it, but a complete circle. Nowhere else is the phenomenon visible, I believe. Like the wanderer of the German legend through the realms of space, I felt satiated with wonders, and with him was ready to cry out, "It is enough! End there is none in the universe of God." Silently and gladly I followed my guide along the awful track back again, and it was no mean sense of relief which I experienced when the hireling recalled me to earth and sense by asking for his fee. "Lo, these are

parts of His ways, . . . but the thunder of His power who can understand." The exact height and distances of the Falls are as follows:
—From Lake Erie to Lake Ontario (36 miles) the total fall is 339 feet, distributed thus:—from Lake Erie to the head of Goat Island (22 miles) 25 feet. From the head of Goat Island to the main Fall (half-a-mile) 50 feet. The perpendicular height of the American Fall, 164 feet; the Canadian or Horse-shoe Fall, 158 feet. From the Falls to the whirlpool (2½ miles), 64 feet, and from thence to Lake Ontario (11 miles), 25 feet. The depth of the water just over the Falls is supposed to be about 20 feet. Dr. Dwight has estimated the quantity of water that goes over the Falls every hour at 100,200,000 tons.

HUNTSVILLE, *Oct.* 5.

We left Niagara on the morning of the 29th ult. for Pelham township, where a local agricultural show was to be held. The seventeen miles' drive took us through some of the best and most fruitful land in Canada. Orchard after orchard and vinery after vinery met our view, and on reaching Fenwick, the village where the show was, other signs of material prosperity appeared. Scores of single and double-horsed "buggies" were standing about, and the street was full of comfortably-dressed farmers, with their wives, sons, and daughters. Conversing freely with these good folks, we gathered a variety of valuable information bearing on our mission. It appeared that while none of the farmers were millionaires, most of them were well off. The township of Pelham contains about 30,000 acres, divided between some three hundred owners. "And what is the average amount of labour employed on these farms?" I asked an intelligent farmer. "Scarcely one paid labourer each," he replied. Here was a revelation touching Canadian farming. Well may the farms everywhere present the labour-starved appearance they do. Each of those three hundred farmers ought to have at least six labourers on his estate, to develop it properly. The fruits of this district were remarkably fine, especially the apples and grapes. From this place we drove over to St. Catherine's, about fourteen miles distant, and took the rail for Hamilton, where another large show was to be held on the morrow. We arrived there towards evening, and found the town full of people. The next day we visited the show, and a splendid affair it was. As we passed through the great buildings where the various products of the country were displayed, astonishment was the most prominent feeling with us. I fancy Englishmen generally have a very imperfect conception of the productiveness and enterprise of Canada. It would have made the fruit merchants of Covent Garden open their eyes to see the magnificent varieties of apples, pears, grapes, melons, &c., which filled the long rows of tables. Not less excellent were the fine roots, such as beet, swedes, turnips, &c., which these Ontario farmers had to display. The horses, too, were a splendid sight. A dozen teams of what they call "general purpose" horses, a sort of cross between our carriage and cart horse, would have been a striking feature even in one of our own great agricultural shows. Canadians evidently pay a good deal of attention to their horses, and few of the farmers are without two or three which would excite admiration in Rotten Row. The light gigs, too, which they drive, are exceedingly elegant, and after riding in them for some hundreds of miles, I can only express the hope that they will soon become general in England. The speed with which a couple of horses will carry you in them from village to village is very great. The horse seems to have nothing behind him. A drive of twenty-five miles without a rest is nothing uncommon. Accompanied by the courteous Mayor of Hamilton, we paid a visit to some of the manufactories of the town. The most striking was that of the Wanzer sewing machines. Here some five hundred hands are employed, and 150,000 dollars are paid yearly in wages. The next morning we returned to Toronto, and had the honour of lunching with the Lieutenant-Governor of the province and a large number of the leading men of the city.

On Friday, the 3rd inst., we started for a tour through this Muskoka district. We rode by the Northern Railway to Washago, about one hundred and twenty miles from Toronto. We then drove along a bad road for some fourteen miles to Gravenhurst, a straggling sort of village—I beg the Canadians' pardon, I should have said town—on the side of the beautiful bay of Muskoka. Here a steamer took us across the bay and up the picturesque river Muskoka some twelve miles to Bracebridge, another Canadian "town." Nothing could exceed the exquisite beauty of the river, which winds like a serpent, and of the autumnal tints of its thickly-wooded sides. The town was disappointing. It consisted of a few straggling wooden buildings, and streets which were mud tracks. The people were ragged and haggard-looking. On the morrow we started on a primitive sort of carriage, drawn by a couple of good stout horses, on a twenty-five miles' drive right through the district to Huntsville. Anything more desolate than that wild track through the forest I cannot conceive. The hundred acres of land given to settlers are a sort of white elephant to the unfortunate recipients. The donation drags them down to the very verge of barbarism. A very intelligent young married lady on board our steamer was returning to her home on one of these grants. Her husband had held a good position in England, but attracted by the representations of a pamphlet on Muskoka, he had thrown up his situation to come out here some seven years ago. He had about a thousand pounds with him, and got a pretty good start. Their life, however, had been one of intense privation and discomfort, and she had been glad to teach in the public school, to eke out their scanty existence. The truth is, none but the hardiest and most persevering men can do any good in these wild regions, and they must lay their account for years of "roughing it." I am driven to the conclusion that if men in England were to work as hard and to live as hard, and to abstain from strong drink, as they do and must do to get on abroad, very few of them would need to leave their old homes. Those poor villagers of my acquaintance in Berkshire, Oxfordshire, and Wiltshire, are rich by comparison with many of these owners of hundreds of acres. They have social comforts and advantages which I look in vain for among the scattered shanties of these Canadian forests. The truth is that the voluntary hardships of Canadian settlers are far greater than any of the involuntary ones which are imposed on English labourers. Take a case in illustration. Just as we were ploughing our way through the last mile or two of mud before reaching the town, we pulled up to speak to a farmer who was standing at his shanty door.

This man was the first settler in this neighbourhood. He told us that when he came nine years ago he had to hold his own against the Indians. One day an old Indian came into his log hut, and, brandishing a formidable club about him, warned him off the place, crying out, "What business have you white man here in our land?" The white man was unarmed, but, by a firm demeanour, managed to get rid of the intruder. There were other Indians close at hand. After years of incessant toil, he has got some forty acres of his land "cleared," by which term it must not be understood that the land is anything like our English fields. They cut the trees down within about three feet of the ground, and when the timber has been either burnt or cut up into "lumber," as the case may be, they call the land "cleared." The final stage of clearing—the removal of the stumps—is extremely hard work, and where it is done extensively you have the best possible proof of the prosperity of the owner. I saw among the implements at Hamilton a machine for the forcible removal of these stumps.

By a reference to the map of the province of Ontario, it will be seen where we are. All around us is the primeval forest. Yonder exquisite sheet of water is a part of the Muskoka river, one of the most picturesque rivers in Canada. Scattered about on the shores of this river are the wooden shanties, the hotel in which I now write, a general store, and a general church, where all sects meet for worship—the Episcopalian in the morning, and some other sects in the other parts of the day. I wish I could convey by my pen an idea of the rare beauty of these interminable forests just now. The deep crimson of the maple trees, the bright yellow of the beech, and the different shades of the ever-green hemlock and balsam, make up a glowing picture. Unfortunately, the phenomenon indicated by the name Muskoka—clear sky— has not been vouchsafed to us during our brief visit. A cloudy sky and a continuous rainfall have been the only welcome accorded us. But even under these adverse circumstances, and traversing a road which would have thrown any English horses, and smashed up any English carriage, my enjoyment was great. I scarcely wonder that amid all their hardships the settlers shrink from the thought of leaving their forest home. After all, it is their own, and they are wholly unfettered by the conventionalisms of society.

LONDON, ONTARIO, *Oct.* 23.

Having received an invitation to visit the celebrated model farm of the Honourable George Brown, the proprietor of the "Globe" newspaper, we left Toronto in the morning, and after a ride of some sixty miles by the Great Western Railway, reached the town of Paris. Here a grand agricultural show was being held. These shows appear to be quite an institution in this flourishing province. We have already attended three or four of them, and from the numbers present, and the extensive display of native products, I take it that their value and importance are pretty well understood by the thriving husbandmen who constitute the main population of the country. After spending an hour or two in the show, and experiencing once more something of the astonishment with which we first beheld a fine display of the fruits, vegetables, and cereals which this prolific soil produces, we accepted an invitation to lunch with the Mayor of the town. This gentleman, who, like every one else here, is a self-made man, is the owner of extensive flour-mills, and we met at his beautiful villa some of the leading men of the district. Each one had his story to tell of early battling with adverse circumstances, and victories won. Mr. Arch found in the foreman of our host an old schoolfellow. He had been in Canada about a quarter of a century. Like every other honest and industrious emigrant, this Barford man had done well. He had bought his own house—a house worth some two hundred pounds—and had perhaps a couple of thousand pounds out at interest. Thus, side by side, were flourishing both master and man. In the afternoon, accompanied by one or two of Mr. Brown's friends, we started for a ten miles' drive to Bow Park. Our road lay through what must, I think, be called the garden of Canada. All the farms appeared to be, unlike those of the Quebec province, which we first went through, in a high state of cultivation. "Yonder," said a gentleman, a brother senator of Mr. Brown, " is my farm of five hundred acres. I bought it thirty years ago, at forty dollars an acre ; and to-day I should want as many pounds an acre for it. Over yonder," at another point, he said, " is a farmer who came out twenty years ago without a cent in his pocket, and now he has two hundred acres of as fine land as is to be found in Canada." Passing by a rather dilapidated-looking farm-house, I remarked on its exceptional appearance. "Yes," said my companion, "the man drinks." This seems to be almost the only impediment to a settler's progress here, but I have previously remarked on the happy rarity of this vice in Canada. We have not seen a single drunkard. One of the causes of this happy state of things is of course the clear, bracing climate; but another is undoubtedly the general discouragement of the habit of drinking strong liquors in the homes of the people. For instance, at the Mayor's table yesterday, iced water and delicious coffee might be had, but neither wine nor ale. At the hotels few drink anything at dinner stronger than tea or coffee, and it is a universal practice to serve these up at the dinner-tables of private individuals.

After a delightful drive of an hour and a half, passing on our way through the flourishing town of Brantford—a place of about 10,000 inhabitants, and the county town of the "Brant" district, so named after the celebrated Indian "Brant"—we reached the point towards which we were tending. Bow Park Farm contains 900 acres, all of it, with the exception of the reservations for ornamental timber, under high cultivation. The farm is nearly surrounded by the grand river, and the soil is alluvial deposit of the most fertile character. The energetic proprietor purchased the estate some few years ago, and has invested a very large capital upon it. Commodious buildings have been erected, one barn alone being about 250 feet long by nearly 50 feet broad. There are about 400 head of high-bred short-horns in the stalls, and no expense is spared in replenishing the stock with the best breeds. The situation of the farm is highly picturesque, and Mr. Brown, like Mr. Mechi, is confident of reaping a pecuniary success, although he has taken to farming as a recreation. It took us the whole morning to go through his extensive range of cattle-sheds, and certainly it was a grand display for so young an establishment. A catalogue of the stock, with full details as to pedigree, &c., was placed in our hands; and at the annual sale, which will be held next week, Mr. Brown expects purchasers from England, Scotland, and all parts of the United States. Not the least interesting part of the exhibition—for such, in truth, the farm is—was the show of Berkshire pigs. It will be gratifying to Berkshire farmers to hear what Mr. Brown said in answer to my inquiry as to whether he had any other breed of pigs. "Of course not," was his reply. "What's the use of having any but the very best?" Some of the thoroughbred shorthorns of Mr. Brown's herd would have won the applause of the severe critics of the great English agricultural shows, and doubtless many of them will yet pass under the review of those gentleman, as Mr. Brown contemplates a considerable exportation of them to the parent country. It would seem that the climate, or feeding, or something else of Canada, vastly improves the breed. Hence the fabulous prices realised occasionally. At a recent sale on this continent, one beast actually fetched forty thousand dollars, and the buyer was an Englishman. Mr. Brown has bought several of his best animals in England, and he will probably get for some of their offspring, from English buyers, treble the original cost of the parents.

It has again and again been asked me by Canadian agriculturists, "Why do not some of your farmers come out here and buy the cleared farms which are always in the market?" The question is easier asked than answered. Certainly it would be the best thing they could do, provided the conditions of success were complied with. One of these is a year's servitude under a good Canadian farmer. Intending emigrants of this class may take this hint as an all-important one. Whenever it is not acted upon, it matters little what may be the capital at command, failure will be the inevitable fate of the farmer. On this there is a marvellous consensus of opinion throughout Canada. "Did you notice that young man?" said our host, as we were passing along his farm buildings. "He is an educated gentleman. He and another Eton lad are with me, to get a thorough knowledge of farming. They live and work with the other men, and are in every respect just like the rest." The "gentleman" was merged in the "working man." As Mr. Brown sententiously remarked, "He's all right." Yes, his success is tolerably sure. By-and-by he will have served his apprenticeship, and then with his two or three thousand pounds capital he will repeat on a small scale the splendid success of his enterprising employer. If a few hundreds of those young gentlemen who are hanging about their fathers' halls in England, trusting to the chapter of accidents, would just follow the example of these young Etonians, they would experience the new sensation of independence.

After a pleasant day among these high-bred cattle, and the picturesque meadows in which they grazed, we took our reluctant departure for the city of London. A two or three hours' rapid ride in a Pullman's car—the perfection of travelling—brought us to this somewhat ambitious namesake of the imperial city three thousand miles away. At the station we found a deputation from the town council waiting to receive us, and after refreshing ourselves with a cup of what seems to be the drink of Canada—tea—we had to yield to the solicitations of the mayor and his colleagues to accompany them round the city. Canadians feel a great satisfaction in showing the proofs of their vigour and enterprise, and these excellent Londoners were no exception to the rule. And certainly they have much to be proud of. At their agricultural show, which we were unfortunately too late to see, no less than 34,000 persons paid for admission, and the exhibition of every kind of agricultural produce was, according to all accounts, remarkable. The further we come west, the greater are the signs of material prosperity. Unlike England, wealth appears to be distributed in almost equal proportions among all classes. Poor people seem to be unknown. In sailing down the Muskoka River, the other day, we had to get our canoe over some rapids, and some young fellows near at hand kindly helped us through our difficulties. We actually had to finesse a little to get them to take a gratuity for their services. It is a somewhat significant fact—and one that English statesmen would do well to ponder—that even the smallest of Canadian villages have their spacious drill-sheds for the use of the volunteers. There is nothing that provokes a smile more readily than a suggestion to a Canadian of danger from invasion. An impression obtains in England that Canada is the weak point of the empire. I am thoroughly convinced that it is a delusion, like many others which are zealously propagated and conscientiously believed respecting the colony in England. It can never be too thoroughly understood by the people and Parliament of Great Britain that this portion of the empire can well take care of itself. Any one who has seen the tens of thousands of lithe, active, intelligent, and well-disciplined men of Canada, and witnessed the innumerable proofs of their patriotism, loyalty, and respect for the parent institutions, will never again have one moment's uneasiness respecting its future. When the unhappy *Trent* business was agitating England, the fear respecting Canada was confined to that side of the Atlantic. Nothing but enthusiasm dwelt in the breasts of these poor relations of the west. The banker left his till, the lawyer his desk, the busy trader closed his ledger and the student his book; and in an incredibly short period the whole country was in a state of preparedness for whatsoever might happen.

I may say, in conclusion, that from hints thrown out by some influential persons, there is good reason for believing that the Ontario Government will co-operate with Mr. Arch in certain very satisfactory arrangements for facilitating the emigration of agricultural labourers to Canada. In all probability, such as would like to come out and settle upon the free grants of Muskoka will have comfortable homes erected for them, a portion of the land cleared for immediate cultivation, and work found by which they will be enabled to get together a little capital to start them upon their farms. Mr. Arch has promised, on condition of these helps being afforded, to bring out next spring a hundred picked men and their families, and see them safely through to their respective lots. An equally satisfactory arrangement has been entered into with the Government immigration agent at Toronto, with reference to the larger body of emigrants who will elect to go into service here for a while. A registry will be kept of applications for such men, giving full particulars as to hours of work, accommodation, &c., certified by the reeve of the township from whence the application comes. Thus, as our mission hither draws near its end, we have good hope that it will be fruitful in beneficial results.

NEW YORK, *Oct.* 31*st*.

A few days spent with Mr. Arch in the neighbourhood of Paris, Ontario, constituted one of the most interesting and satisfactory episodes of his Canadian mission. He there found several of his old Barford acquaintances, men with whom he had worked, and with whom as boys he had gone to the village school. For periods varying from five to thirty years these men have been in Canada. They were, therefore, in a position to give good, sound information on the question of agricultural labourers' emigration thither. No possible motive for misrepresentation could influence them. And what was their unanimous verdict? In the highest degree favourable. By their words, by their social positions, by the comfort of their families, they said most emphatically to their English brother labourer, "Advise your clients, the farm labourers of England, to come out to this labour-starved country. They cannot fail, if honest, sober, and industrious, not merely of getting a good living as servants, but of becoming proprietors of the soil. Here are we, owners of good farms, possessors of well-built houses, with money out at interest, with children being well educated, and thereby fitted to occupy the best positions in the State. Had we remained at home, as others did, where should we have been but where they are—pacing wearily the social treadmill, toiling hard from week to week and from year to year for inadequate wages, ever on the edge of the precipice of want, the pauperised hangers-on of the poor-law board, or discontented agitators whom landlords, priests, and farmers might combine to denounce as the nuisances of their respective parishes." Sitting at their well-spread tables, and looking out upon their well-stocked farms, Mr. Arch has been compelled to acknowledge that the evidence is complete. The land which he had come to spy is emphatically "a good land." Thus terminated our Canadian mission, and as I notice that American papers have caught up very greedily the somewhat disparaging comments which I felt bound to send you on the agricultural prospects of the eastern province of Canada, I hope they will as readily reproduce my equally emphatic testimony of another character respecting the western province.

On the 23rd we left for Ottawa, having received a telegram from Mr. Lowe, the Secretary for the Agricultural Department, intimating that his chief, the Hon. Mr. Pope, particularly desired a final interview before we returned to England. As it was the day for the reassembling of Parliament, our train was crammed with senators and M.P.s, and the excitement of the past three months respecting the Pacific Railway business was evidently approaching its crisis. No other talk was heard during the long, tedious journey, but the respective merits and demerits of the Government. The unfortunate Premier at Ottawa, Sir John A. Macdonald, appears to be about the best-abused man in Canada just now. I have listened to strong language in England in times of political excitement, but never have I heard, even from the rowdies of large cities, such wholesale abuse poured forth upon a statesman, as I heard heaped upon the head of the Canadian Premier. It is not, of course, a part of our mission to enter into such questions, but it is impossible for an Englishman to feel otherwise than deeply interested in matters which are in their bearings imperial rather than provincial. No one can come into contact with Sir John A. Macdonald without feeling that he is in the presence of no ordinary man. His severest critics freely admit that his services to the Dominion have been neither few nor small. Hence the toughness of the task assigned to Mr. Mackenzie, the leader of the Opposition, of driving him from power.

On the morning of the 24th we had a long interview with Mr. Pope, and the result was in the highest degree satisfactory. He was fully prepared to co-operate with the National Agricultural Labourers' Union in an extensive scheme of colonial emigration. He would have been glad, indeed, if Mr. Arch could have accepted an appointment in England from the Government of the Dominion. A salary much larger than anything he had ever received from the Union would have been readily offered him. Of course this was out of the question. Mr. Arch is no office-seeker. His one aim in life is to raise and benefit others—not to exalt himself. His emphatic *non-possumus*, therefore, at once dismissed this tempting offer of the minister. Not the least important of the points conceded to Mr. Arch was the offer of Mr. Pope to take the signature either of himself or the secretary of the Union, in lieu of a clergyman's or magistrate's, to the forms of application for assisted passages. A book of warrants for such assistance will be forwarded to the Leamington office, and thus members of the Union will be henceforth entitled, on the recommendation of their president or secretary, to a grant towards the passage to Canada, amounting to about forty-five shillings each adult. Mr. Pope was pleased to express the great satisfaction with which he had watched our progress through the Dominion. Our plan of avoiding all noisy demonstrations, and keeping closely to our work of observation and investigation, seemed to have afforded him immense satisfaction, and his confidence in the judgment and good practical common sense of Mr. Arch was evidently all that even English landlords and farmers could have desired.

In saying a few last words touching this Canadian mission, I cannot resist the temptation of directing attention once more to the uniform courtesy accorded to the representatives of the English agricultural labourers by the Government and people of the Canadian Dominion. We brought no credentials to their shores. No potent line from Downing-street preceded us. A few private letters of introduction were all that we had. And yet from the moment we embarked at Quebec to the hour when we last week stepped into the New York train at Prescott junction, we received every attention. Had we been honoured with the *imprimatur* of the British Government we could not have received more. Our hotel bills were discharged. Free passes over the railroads were given to us. Carriages were placed at our disposal to enable us to penetrate those remote regions where the iron rails had not yet been laid. Efficient guides were deputed to facilitate our researches. The municipal authorities of the various towns along our route to Ontario extended to us their generous hospitalities, and last, though not least, the press, with scarcely one exception, bade us a very hearty " God speed !" Of course cynics reply to all this that the Canadians know what they are about. Undoubtedly they do, but I take it that even this assumed self-interest in the matter does not lessen the respect and gratitude which are due to our colonists for their cordial welcome of the man whom many of his own countrymen affect to despise.

And I would respectfully submit for the consideration of all whom it may concern, whether it is not pretty well time for other than Canadian authorities to treat with something better than cold contempt a man and a work whose aim it is to elevate an important section of the British community. When will English landlords learn the value of that which Canadian landlords are so anxious to procure? Is not the measure of the pains taken by the latter to allure the labourers to their fields the true measure of their worth? Mr. Arch has repeatedly said that Emigration is his "strange work." He was long in direct opposition to it. Could he but see his brother labourers emancipated from the wretched thraldom in which he sees them in England, and possessing a fair chance of earning an independence, in the shape of an acre or two of land to their respective homes, held direct from the landlord, and with proper security of tenure, Canada and the United States might offer their inducements in vain for him. He would elect to see his people settling down peacefully on their native soil, and enjoying with himself the unique advantages of England.

The unsettled condition of things in New York, together with the lateness of the season, have determined us to defer for the present an investigation of the States. We therefore purpose returning home by the White Star line steamer the *Republic*, on Saturday, the 8th prox. On Monday we go to Boston, at the urgent solicitation of a deputation of citizens, who are bent on giving Mr. Arch a welcome such as they think he deserves. This will bring our mission to a close.

ON BOARD THE STEAMER "REPUBLIC,"
Nov. 17.

The protracted stay of Mr. Arch in Canada, rendered necessary by the immense area to be traversed, left him little time in which to see the States. He therefore resolved, with the full concurrence of his colleague, to abandon the idea of anything like a systematic survey for the present. Another and a more convenient season must be chosen in order that the vast territories of the Republic may be done equal justice to with Canada. As regards British America, there need be no hesitation in stating that Mr. Arch's opinion respecting it is on the whole exceedingly favourable, although he will require, on the part of both the Government and people of Canada, some very important modifications of their arrangements before he can commit himself to an unqualified recommendation of the colony as a field for emigration. The Canadians are for the most part self-made men, and are great believers in what they call "roughing it." Their unfailing argument against anything in the shape of provision for the comfort of coming emigrants is their own experience. "We came out with next to nothing," say they, "and with no one to look after us, and here we are, well-to-do, prosperous men. Bid your clients come and do likewise." Mr. Arch's reply is that the relative circumstances of the different times demand different treatment. The civilisation of to-day is widely different from that of half-a-century ago. The roughing of these men in that day was but a trifling change from their home life. Now the poorest peasant has, as the necessaries of his life, what would then have been deemed luxuries. To expect, therefore, similar disregard of social comfort now, as then, is unreasonable. Mr. Arch must have houses found for his fellow-labourers if they are to become tillers of Canadian soil; and if the Dominion is desirous of an English agriculturist settlement on its boundless prairies and forests, it must offer them the conditions of something like civilisation. It is highly satisfactory to be able to say that both the Dominion Government and the Governments of the Provinces fully recognise the importance of this opportunity of getting what their country so sorely needs, and there is no doubt of their fulfilling his utmost wishes. He brought from Ottawa the amplest guarantees of this, and although a change of Government has taken place, there will be no change in emigration policy, as all parties are equally alive to its supreme importance.

Mr. Arch found himself the subject of intense curiosity in the States. The newspapers had heralded his approach, and were loud in his praise. Letters from all parts poured in upon him as soon as his arrival was announced, and the apartment at his hotel was occupied from morning till night by persistent interviewers. The day after his arrival in New York nearly every paper had its pen and ink description of the Warwickshire labourer. His height, shape, hair, clothes, all were described with amusing minuteness. I am afraid the New Yorkers will not be best pleased with the Labour Reformer's attitude towards them. Mr. Arch did not understand their ways. It appears that two months ago it was heralded all over the city that he had arrived from England, and would address a monster meeting of working men in the Cooper's Institute. The newspapers enlarged on the affair, and at the appointed hour thousands were assembled. I believe an entrance fee was charged. "And who got up the infamous hoax?" asked the straightforward Warwickshire man of his informant. "Oh, the Central Working Men's Union, or some such organisation," answered he. "Then I'll have nothing whatever to do with them," replied Mr. Arch. To this resolution he firmly adhered. It was in vain that one after another came, beseeching him to address them. "No," said he, "you have acted dishonourably towards the public, and, whatever may be your American customs, we in England only know two things, right and wrong. You had no authority whatever for publishing my name as a speaker at your meeting. You did not even know where I was, and yet you deliberately advertise me all over the city, and so associate my name with an imposition. I wish you good-morning, gentlemen; I'll have nothing whatever to do with you."

A pressing invitation from Boston to attend a working men's demonstration met with more favour. Although, however, the New York Trades Unionists thus got the cold shoulder from Mr. Arch, his co-religionists were more successful. On Sunday, the 2nd instant, he preached to a large congregation in one of the Brooklyn Primitive Methodist Chapels. The number of letters arriving daily during our stay in New York testified to the great interest felt in the English agricultural labourers' movement throughout the United States. On the 3rd instant Mr. Arch started for Boston, and on the 5th, accompanied by Mr. Wendell Phillips and other prominent politicians, he made a tour of the city. In the evening the promised reception took place in the Fancuil Hall. It was certainly a splendid demonstration. Mr. Arch, as is usual on all important occasions, was fully equal to the necessities of the hour. His speech took every one by surprise. Those who had come, like Mr. Wendell Phillips and General B. Butler, to patronise the rejected of English squiredom, were lost in admiration of the perfect self-command and ease of expression displayed by the English farm-labourer. More than one influential gentleman by my side on the platform assured me that they had never heard a more effective hour's speech in their lives. When, in illustration of the animus against which he had to contend, Mr. Arch referred to the bankrupt prosecution of Chipping Norton, cries of " Shame " were heard all around. Those clerical justices and the manly prosecutor may congratulate themselves on a considerable augmentation of the area of their fame. And it is interesting to have to record that the applause which Mr. Arch received was not the gratitude of an England-hating assembly for abuse of her fair fame. This he distinctly repudiated, and when his companion, Mr. A. Clayden, referred, in a short speech towards the close of the proceedings, to the fact that it was no part of their mission to defame their country, which they loved with an affection vastly increased by its comparison with other countries, cheers as hearty as they were sincere greeted the declaration.

The day following the Mayor of the city gave us an invitation to dine, having previously taken us to see the poet Longfellow, and other celebrities and attractions of the city. At this final celebration of the time-honoured institution of civic hospitality, a very different circle of guests were invited to those whom we met at our first complimentary banquet at Quebec. The Mayor had invited some half-dozen representative working men to meet their friend. It was all the same to Mr. Arch. As he told the Bostonians, in a playful allusion to the petty sneers of some at home about his being feasted by the rich, that as an Englishman he had a profound respect for a good dinner, and, whether it was in the cottage of a labourer or the palace of a king, if he found a hearty welcome, he was equally prepared for the feast.

At 5.30 p.m. we took the train for the point where the evening steamer for New York would start from some two hours later. In this we steamed through the night down the magnificent Fall River, and in the morning took possession of the cabin generously placed at our disposal by the owners of the steamship *Republic*, one of the very fine vessels of the White Star Line.

Thus terminates this mission from the National Agricultural Labourers' Union. So far as it has been prosecuted, Mr. Arch feels satisfied with its success; and if, as he is hoping, he has the opportunity of doing next year in the United States what he has done in Canada, he will certainly have gained sufficient data for emigration purposes, to enable him to act as an efficient guide to all who, from choice or necessity, contemplate a removal from the parent country.

www.ingramcontent.com/pod-product-compliance
Lightning Source LLC
Chambersburg PA
CBHW031743230426
43669CB00007B/464